Springer Tracts in Modern Physics
Volume 151

Springer
Berlin
Heidelberg
New York
Barcelona
Hong Kong
London
Milan
Paris
Singapore
Tokyo

Springer Tracts in Modern Physics

Springer Tracts in Modern Physics provides comprehensive and critical reviews of topics of current interest in physics. The following fields are emphasized: elementary particle physics, solid-state physics, complex systems, and fundamental astrophysics.
Suitable reviews of other fields can also be accepted. The editors encourage prospective authors to correspond with them in advance of submitting an article. For reviews of topics belonging to the above mentioned fields, they should address the responsible editor, otherwise the managing editor.
See also http://www.springer.de/phys/books/stmp.html

Hansjörg Donnerberg

Atomic Simulation of Electrooptic and Magnetooptic Oxide Materials

With 45 Figures and 40 Tables

Springer

Dr. Hansjörg Donnerberg
Universität Osnabrück
Fachbereich Physik
D-49069 Osnabrück
Email: hdonner@physik.uni-osnabrueck.de

Physics and Astronomy Classification Scheme (PACS):
61.72.-y, 71.10.+x, 71.55.-i, 78.20.-e

ISSN 0081-3869
ISBN 3-540-65111-X Springer-Verlag Berlin Heidelberg New York

Library of Congress Cataloging-in-Publication Data.

Donnerberg, Hansjörg, 1959- . Atomic simulation of electrooptic and magnetooptic oxide materials/Hansjörg Donnerberg. p.cm. – (Springer tracts in modern physics, 151). ISBN 3-540-65111-X (hc.: alk. paper). 1. Semiconductors-Materials-Computer simulation. 2. Metal oxides. 3. Electrooptics-Materials. 4. Magnetooptics-Materials. I. Title. II. Series. QC1.S797 vol. 151 [QC611.8.M4] 539 s-dc21 [537.6'22] 98-46730

Typesetting: Camera-ready copy by the author using a Springer TEX macro package
Cover design: *design & production* GmbH, Heidelberg

SPIN: 10688541 56/3144 - 5 4 3 2 1 0 – Printed on acid-free paper

Preface

The present compendium reviews modern state-of-the-art computer simulation methodologies commonly used to model relevant defect properties of technologically important complex oxides on an atomistic level. It provides a broad discussion of most of the results obtained so far.

This book will be invaluable for all readers interested in acquiring a detailed but quick overview of this research field. The text is written on a level such that researchers, lecturers and graduate students may benefit from the discussions presented.

Ultimately this book has become possible thanks to many important collaborations and helpful discussions with experts in this field. In particular I am indebted to Prof. O. F. Schirmer, Prof. C. R. A. Catlow, Prof. R. H. Bartram and Prof. E. A. Kotomin.

Osnabrück, October 1998 *Hansjörg Donnerberg*

Contents

1. Introduction

In many instances, insulating oxide materials behave semi-ionically, with bonding properties ranging between ordinary ionic systems (such as alkali halides) and semiconducting crystals, which are affected by significant covalency contributions. Structural and physico-chemical modifications give rise to a broad class of solid materials with many interesting prospects for technological applications. This book is devoted to reviewing recent atomistic simulations of insulating complex electro- and magnetooptic oxides. The possible prospects of these materials include the development of optical device elements allowing fast and efficient transport and processing of information. Photorefractive crystals form an important subset of electrooptic materials; they facilitate the transformation of an incident light pattern (containing information) into a volume refractive index pattern and, thus, represent useful holographic storage media. Magnetooptic materials such as suitable garnets permit the development of non-reciprocal waveguide components. Further details of the technological prospects are given in [1, 2] and [3, 4, 5] for electro- and magnetooptic materials, respectively.

The fundamental bonding properties of insulating oxides rely to a considerable extent on the electron affinity of oxygen. The accommodation of two excess electrons – leading to O^{2-} anions – allows the formation of a stable noble gas configuration, being isoelectronic to neon. However, the doubly negatively charged oxygen anions are unstable as free ions, and, thus, their stabilization requires a crystalline Madelung potential and, possibly, additional covalency contributions. There exist, besides perfectly ionic oxides (like MgO), a variety of insulating oxides which possess open crystal structures and are affected by appreciable covalency contributions. Their crystal structure often deviates from that of simple cubic systems. The growing significance of the electronic structure (due to covalency effects) in these oxides can lead to pronounced dielectric material properties. Most of the oxides discussed here are, indeed, ferroelectric. They are non-centrosymmetric and have pronounced electrooptic material constants. Important magnetic properties, on the other hand, arise due to cooperative effects involving paramagnetic transition metal cations. Technologically important material properties are influenced or even made possible by incorporated defects. Most striking is the example of photorefractive oxides. The operation of photorefractive effects is accomplished through deep impurity levels in the band gap of the materials. Typically, the band gap in semi-ionic oxides ranges between 3 and 4 eV, enabling processing technologies in the visible regime. A prerequisite for the possible controlled design of the desired material properties is an understanding of their microscopic origin. This requires detailed knowledge of perfect lattice properties and, importantly, the various relevant features of lattice imperfections. Therefore it appears to be mandatory to investigate the possible chemical origin of defects and their structural and electronic properties. Potential simulations and electronic structure calculations provide helpful theoretical tools that provide the knowledge of the relevant material properties on a microscopic level. The compilation of corresponding results represents the major goal of the present contribution.

A few more introductory remarks are useful in order to sketch the broad variation of properties of insulating oxides. Oxides enable large variations of possible cation charge states. The ionically bonded magnesium oxide possesses a cubic close-packed NaCl structure. The cations are octahedrally coordinated, and this is frequently observed in other oxides, too. Large variations, on the other hand, occur for anion coordination. The corresponding coordination numbers range between CN=6 in MgO and CN=2 in the less ionic tungsten trioxide WO_3. Because CN=6 in MgO the charge state of the octahedrally coordinated magnesium is +2. A reduction of the anionic coordination number leads, first, to an opening of the crystal structure, and, second, to the occurrence of higher cation charge states. Whereas the formal cation charge state is +2 in the densely packed MgO it increases to +6 in WO_3. The basic WO_6 octahedra are corner sharing, thus leaving much space for interstitial sites. At the same time the stabilizing effect of the Madelung potential can decrease in favour of stabilization due to covalent charge transfer between anions and cations. Covalency contributions also allow a partial reduction of the large formal charge states. But significantly, in many situations structural characterizations based on ion size considerations remain reliable. Intermediate cation charge states between +2 and +6 can be easily achieved either by modifying the oxygen coordination number or by incorporation of additional cations at interstitial sites. Crystal structures can be built up involving corner-, edge- or face-sharing metal oxygen octahedra. Corresponding oxides are B_2O_5 (+5), BO_2 (+4) and B_2O_3 (+3). Alternatively, the open structure of the idealized cubic BO_3 oxides allows the accommodation of variable amounts of additional cations A. This leads to the occurrence of solid solutions A_xBO_3 ($x \leq 1$). These bronze-type systems can show pronounced n-type semiconducting properties. In this fashion one may understand the formation of insulating oxide perovskites ABO_3 ($x = 1$) and also of strontium barium niobate SBN (Chap. 6) and related materials [6]. These crystals represent important electrooptic oxides. In perovskites the introduction of divalent A cations reduces the charge state of B cations to +4, but the corner sharing of oxygen octahedra is retained at the same time. An important example is given by barium titanate $BaTiO_3$ (Chap. 3). Monovalent A cations relate to pentavalent B cations. Potassium niobate and tantalate ($KNbO_3$ and $KTaO_3$, Chap. 4) belong to this category, as does lithium niobate $LiNbO_3$ (Chap. 5). However, the small size of the lithium cations leads to a distorted perovskite structure analogous to corundum. Finally, the high-T_c oxides should also be mentioned in this context, since their crystal structure closely resembles the perovskite structure.

Appropriate cations allowing of many different formal charge states are provided by transition-metal atoms. Due to their natural abundance these are often found as cation partners in higher-valent oxides. In electrooptic materials (like $BaTiO_3$, $KNbO_3$, $KTaO_3$, $LiNbO_3$ and SBN) the B cations are characterized by d^0-type electronic configurations. Partially filled d-subshells, on the other hand, lead to the occurrence of magnetic moments giving rise to cooperative magnetic effects such as ferro-, antiferro- or ferrimagnetism. Of particular technological interest are magnetic garnets, of which the ferrimagnetic yttrium iron garnet (YIG, Chap. 7) represents an important member. The garnet structure is built up of corner-sharing FeO_6 octahedra and additional FeO_4 tetrahedra. The iron cations occur as trivalent paramagnetic species with $S = 5/2$. The structure is stabilized upon the incorporation of diamagnetic trivalent yttrium cations which fill up the interstitial cation sites present in this open crystal structure.

The various modifications of possible crystal structures and cation charge states in semi-ionic oxides open the way for a range of materials with different material properties. Optimistically one may think of the possibility of molecular engineering which aims at the optimization of material properties. The discussion presented above suggests that a tailoring of macroscopic material properties seems to be possible upon admixing different cation species. This behaviour can also be expected even for small impurity concentrations. Notably, by analogy with the variety of possible cation charge states in perfectly grown oxides, the materials also allow the solution of many impurity cations within a range of stable charge states. Their stabilization is accomplished to some extent by suitable hybridizations between cations and oxygen anions. Originally, this scenario was suggested by Haldane and Anderson [7] to be operative in semiconductors. Additional contributions are due to defect-induced lattice relaxations which typically occur in ionic materials.

Central aspects of theoretical defect studies include the formulation of favourable defect formation reactions (e.g. solution modes of impurities specifying their lattice sites and possible charge-compensating defects) and calculating the electronic properties (e.g. ground and excited defect levels) of dominant defect structures. Obviously, both aspects are highly dependent on each other. As has been discussed, the specified oxides are intermediate between ionic and covalent materials. Consequently, the charge compensation of impurities is not restricted to electrons and holes (as is mostly the case in ordinary semiconductors), but involves a rich variety of structural defects which are already known for proper ionic crystals. Similarly, defect-induced lattice relaxations should be taken into account when studying the formation and properties of defects. Whereas potential simulations are successful in elucidating the defect chemistry of materials, the local electronic defect properties can best be investigated on the basis of embedded cluster calculations.

Impurity defects are able to modify the electrical, optical and, evidently, magnetic properties of the respective oxides. The charge state and site of incorporation of impurities determine their donor/acceptor behaviour. In electrooptic materials the optical properties are particularly modified by impurity-related charge transfer transitions leading to broad absorption and emission bands. Intra-ionic crystal field transitions, which leave the impurity charge state unchanged, are of less importance in this respect, but they become relevant in the development of efficient solid state lasers, for instance. The properties of free charge carriers determine the electrical conductivity of oxides. These carriers may be electrons or holes depending on the actual position of the Fermi level. Due to pronounced electron–lattice coupling the formation of small polarons may represent the most favourable free carrier states in semi-ionic materials. This is particularly true for d-type conduction band electrons owing to the comparatively small band dispersion. Free charge carriers are also important in photorefractive materials, where they appear upon illumination due to charge transfer transitions. In perovskite oxides one frequently observes that holes define the dominant charge carriers. During their transport holes are trapped at suitable acceptor defects which effectively aid their localization. Trapped holes are known to influence the photorefractive material properties. Chapter 3 includes reports of some recent simulations of trapped hole centres in $BaTiO_3$. In some cases one even expects the formation of small hole bipolarons (Chap. 3). The stabilization of these peroxy type species is due to covalency and defect-induced lattice relaxations. Possibly such bipolarons could also account for high-T_c superconductivity, following the suggestions of Alexandrov and Mott [8]. Electronic counterparts become important if the Fermi level is sufficiently high. Corresponding species have been established to exist in Ti_4O_7, for example, and also in lithium niobate. In these cases the electrons are located at neighbouring cation sites. Analogous to hole bipolarons, the stabilization of electron bipolarons invokes covalency effects and lattice relaxation. Electron bipolarons may also exist in other oxides such as perovskites (see Chap. 3). A review of the evidence for electron bipolarons has been given by Schlenker [9].

Finally, impurity defects may act as polarizing lattice perturbations enhancing the potential electrooptic applications of the materials. This is particularly true for $KTaO_3$, which remains cubic and therefore non-polar down to low temperatures unless there are effective polarizing defects. A characterization of such defects is given in Chap. 4.

Besides impurity cations there may also be a strong influence due to intrinsic defect structures. In ionically bonded oxides the formation of isolated intrinsic defects is rather unlikely unless they appear as charge compensators of impurities. In comparison with alkali halides, one must expect larger lattice energies and intrinsic defect formation energies due to the higher ion charge states in oxides. For example, the formation of a Schottky pair requires 2.2 eV in KCl but 7.5 eV in MgO [10]. This trend can be observed in many complex oxide crystals, too. However, purely intrinsic defect formations may become important upon massive aggregation of defects leading to extended defect clusters. In this fashion the observed non-stoichiometry in many transition metal monoxides can be understood (e.g. see [11, 12]). For example, the appreciable cation non-stoichiometry observed in 3d-transition-metal monoxides, $M_{1-x}O$, has been attributed to complex defect aggregations. Further examples refer to shear planes which may form in higher-valent oxides (e.g. TiO_2) upon reduction. Pronounced oxygen deficiency can by accommodated by increasing the number of shared oxygen anions between neighbouring metal oxygen octahedra, thereby reducing the metal charge states close to the interface (see also above). There are also indications of extended defect clusters accommodating the Li_2O non-stoichiometry in $LiNbO_3$ (Chap. 5). In $LiNbO_3$ the intrinsic defect structure strongly influences impurity solution modes and modifies material properties.

The following chapters compile the results of atomistic simulations of electro- and magnetooptic oxides. The materials considered include the ferroelectrics $BaTiO_3$, $KNbO_3$, $KTaO_3$, $LiNbO_3$ and SBN as well as the ferrimagnetic YIG. For these oxides extensive work has been reported so far. Particular attention will be paid to the fundamental investigation of favourable defect structures in these oxides. It is noted that in ferroelectric perovskite materials all structural deviations from cubic symmetry are small. Therefore, defect simulations can be based with sufficient accuracy on the ideal cubic structure of these oxides, because defect-induced perturbations are likely to be predominant.

Before discussing the details of the respective material properties, Chap.2 reviews the most important features of the different existing simulation methodologies. The many modelling studies that I have performed during recent years employ potential simulations and embedded cluster calculations. These techniques are particularly useful in investigations of perfect and defective semi-ionic materials because they take defect-induced lattice relaxations fully into account. Publications which have strongly influenced the content of this habilitation thesis are listed at the end of the book.

2. The Scope of Theoretical Methods

In what follows, the emphasis is on theoretical approaches towards modelling of perfect and defective electro- and magnetooptic oxide materials. Principally the invoked methodologies employ *ab initio* calculations concentrating on the electronic structure of the materials and potential simulations referring to the subsystem of the nuclei.

Basically the Helmholtz free energy F of the whole crystal,

$$F = -kT \cdot \ln\{\mathrm{Tr}(\exp(-H/kT))\}\,, \tag{2.1}$$

completely describes the equilibrium states of crystals at a given temperature T. Its calculation involves a minimization procedure with respect to all nuclear and electronic degrees of freedom. H is the total Hamiltonian of the system:

$$H = T_\mathrm{n} + V_\mathrm{nn} + H_\mathrm{e} \tag{2.2}$$

$$H_\mathrm{e} = T_\mathrm{e} + V_\mathrm{ee} + V_\mathrm{ne}\,. \tag{2.3}$$

Strictly, the evaluation of (2.1) requires us to calculate the trace with respect to complete bases of the nuclear and electronic subspaces. However, this is an unfeasible task in practice. Instead, particularly for non-metallic systems, only one electronic state (e.g. the ground state) is effectively taken into account. This corresponds to employing the Born–Oppenheimer approximation, which allows separate consideration of the electronic and nuclear subsystems. One first solves the electronic problem, i.e. formally

$$E_0(\mathcal{R}) = \min_{\Psi}\langle H_\mathrm{e}\rangle_\Psi\,. \tag{2.4}$$

$\mathcal{R}$ denotes the total nuclear configuration of the crystal and Ψ the many-electron wavefunction. This variational problem of a many-body electronic system is usually solved by standard Hartree–Fock (HF) methods or by applying density functional theory (DFT). Only for small systems of molecular dimensions may HF treatments be extended by configuration interaction (CI) expansions to account for electronic correlations. In Sect. 2.1 we will briefly describe the state-of-the-art techniques used to model the electronic properties of perfect and defective crystals.

Having found a suitable solution for the electronic subsystem, the next step considers the subsystem of nuclei:

$$H_{\mathrm{n}} = T_{\mathrm{n}} + V_{\mathrm{nn}} + E_0(\mathcal{R}) =: T_{\mathrm{n}} + V_{\mathrm{eff}}(\mathcal{R})\,. \tag{2.5}$$

In principle, the investigation of this Hamiltonian allows us to obtain information on

- equilibrium geometries,
- perfect lattice properties (e.g. phonons, elastic constants and bulk moduli),
- instabilities with respect to ferroelectric phase transitions (FE-PT) and
- ionic aspects of defect formation (formation enthalpies, agglomeration of defects, defect-induced lattice relaxation etc.).

In most cases H_{n} is treated classically, which approximately replaces the quantum mechanical treatment in principle. In Sect. 2.2 we will briefly consider approaches to obtain and use effective potentials V_{eff} in perfect and defective lattice simulations. It is noted in this context, that the approach of Car and Parrinello [13] unifies the above treatments of both subsystems in the framework of molecular dynamics (MD). The method is also based on the validity of the Born–Oppenheimer approximation. One introduces supercells satisfying periodic boundary conditions and performs a dynamical simulation with respect to the nuclear coordinates of ions or atoms in a supercell (physically true dynamics) and to the electronic variational parameters (e.g. one-electron orbitals ψ_i) and possible external constraints α_i defining a fictitious dynamics. Formally, the relevant Lagrangian can be written as

$$\mathcal{L} = T_{\{\psi_i\}} + T_{\{\alpha_i\}} + \sum_j \frac{1}{2} M_j \dot{\mathbf{R}}_j^2 - E[\{\mathbf{R}_i\}, \{\psi_i\}, \{\alpha_i\}]\,. \tag{2.6}$$

$T_{\{\psi_i\}}$ and $T_{\{\alpha_i\}}$ denote the kinetic energies related to the fictitious dynamics and $E[.]$ is the total energy of the investigated system corresponding to the Hartree–Fock approximation (which may be extended to configuration interaction schemes) or to density functional theory. Approaching equilibrium, the equations of motion for the electronic variational parameters reduce to the effective one-electron Hartree–Fock or Kohn–Sham equations (see Sect. 2.1). The numerical efforts of MD simulations increase significantly with the degrees of freedom of the investigated system (electronic structure, number of ions in a supercell). This explains the almost exclusive use of the approach in combination with pseudopotentials and density functional theory [13, 14]. The Car–Parrinello method can be employed in two entirely different ways, i.e. to perform energy minimizations on the basis of simulated annealing or to do genuine molecular-dynamics simulations.

Though the Car–Parrinello method will be used increasingly with growing computer capacity, it is at present ineffective even for simulations of FE-PT in oxide perovskites ABO_3 using the smallest possible supercell (i.e. primitive unit cells containing five atoms). Therefore, in practice potential simulation methods are employed (Sect. 2.2). Problems involving defects are significantly more demanding, as this would require the use of very large supercells in order to avoid artificial defect–defect interactions. This may inhibit full-electron

MD simulations of many defective complex materials. Recent applications of the Car–Parrinello method have addressed the calculation of the vacancy formation energy in aluminium [15].

Fortunately, in most situations of practical interest the quasi-static approach used in (2.4)–(2.5) has been found to characterize the electronic defect structure reliably as well as the atomistic aspects of defect formations.

2.1 Electronic Structure Calculations

The variational condition (2.4) can be investigated by employing the Hartree–Fock approximation or using density functional theoretical methods. The HF approach corresponds to considering the electrons of the system as essentially being independent of each other. The many-electron wavefunction in (2.4) is taken as a product of one-electron spin orbitals which is further antisymmetrized with respect to particle permutations (Slater determinants). The variational procedure yields the well-known Schrödinger-type HF one-electron equations,

$$\begin{aligned}&\left(-\frac{1}{2}\Delta-\sum_i\frac{Z_i}{|\boldsymbol{R}_i-\boldsymbol{r}|}+\int\frac{\varrho(\boldsymbol{r}')}{|\boldsymbol{r}'-\boldsymbol{r}|}\mathrm{d}^3r'\right.\\&\left.-\int\frac{\varrho_x(\boldsymbol{r},\boldsymbol{r}';k)}{|\boldsymbol{r}'-\boldsymbol{r}|}\mathrm{d}^3r'\right)\psi_k(\boldsymbol{r})=\varepsilon_k\psi_k(\boldsymbol{r})\,,\end{aligned}\tag{2.7}$$

or in abbreviated notation

$$\left(-\frac{1}{2}\Delta+\mathcal{V}_{\mathrm{HF},k}(\boldsymbol{r})\right)\psi_k(\boldsymbol{r})=\varepsilon_k\psi_k(\boldsymbol{r})\,.\tag{2.8}$$

The total electron density is given by

$$\varrho(\boldsymbol{r})=\sum_n^{\text{occupied}}|\psi_n(\boldsymbol{r})|^2\,.\tag{2.9}$$

The effective one-electron Hamiltonian in (2.7) is called the Fock operator.

By expanding the one-electron orbitals with suitable basis functions the set of integro-differential Hartree–Fock equations is transformed into a (nonlinear) matrix equation for the expansion coefficients. In periodic systems (perfect crystals, see also Sect. 2.1.1) these basis functions may be either (possibly augmented) plane waves or localized functions (atomic orbitals, Slater- or Gaussian-type orbitals [16, 17]). Localized basis functions are also appropriate for molecules or extended non-periodic systems.

According to (2.7) each electron moves in the field of the nuclei and in the averaged effective field of all the other electrons (mean field approach); the total effective one-electron potential denoted by $\mathcal{V}_{\mathrm{HF},k}$ in (2.8) depends also on the orbital ψ_k, which is to be determined. The use of Slater determinants results in the (non-local) exchange potential energy term given by the (ψ_k-dependent) exchange density $\varrho_x(\boldsymbol{r}, \boldsymbol{r}'; k)$. This exchange density (Fermi hole) describes the density reduction around a specified electron located at $\boldsymbol{r}$ and is caused by all electrons having the same spin alignment. The normalization of the exchange density,

$$\int \varrho_x(\boldsymbol{r}, \boldsymbol{r}'; k) \mathrm{d}^3 r' = 1 \,, \tag{2.10}$$

and the correct asymptotic behaviour of the exchange potential, i.e.

$$\int \frac{\varrho_x(\boldsymbol{r}, \boldsymbol{r}'; k)}{|\boldsymbol{r}' - \boldsymbol{r}|} \mathrm{d}^3 r' \sim \frac{1}{|\boldsymbol{r}|} \qquad (|\boldsymbol{r}| \to \infty) \,, \tag{2.11}$$

guarantee the exact cancelation of the artificial self-interaction which is present in the Hartree term. The use of an averaged exchange potential which is then experienced by all electrons refers to the Hartree–Fock–Slater approximation [18].

It is an advantage of the Hartree–Fock approximation that it allows us to calculate both the ground and excited electronic states. This is true because the HF one-particle equations are derived from the stationary state condition $\delta E = 0$ without assuming any ground state conditions. The evaluation of energy separations between two different states should be performed on the basis of two independent SCF calculations (ΔSCF method). The use of one-particle energies ε_k, on the other hand, is of limited value, since orbital relaxations resulting from the electronic redistributions are totally neglected in this way. In particular, localized orbitals with atomic or molecular extensions are affected by such relaxation effects (see also Sect. 2.1.2). The disadvantages of Hartree–Fock procedures are, first, the pronounced consumption of computer resources related to the huge number of two-electron integrals describing interelectronic Coulomb and exchange effects (for example, if the HF equations are expanded with L basis functions the number of two-electron matrix elements increases as L^4). Second, HF theory totally neglects electron correlations; the correlation energy is defined as the total energy difference between the exact non-relativistic energy and the HF energy.

There are several possibilities to reduce the required computer resources of full HF calculations (see references [16, 19] for details):

- Use of ZDO (zero differential overlap)-type procedures by means of which all or part of the two-electron exchange integrals are set equal to zero. A general feature of all these schemes is that the number of non-vanishing two-electron matrix elements is proportional to L^2. However, with this traditional approach important interelectronic interactions can be lost.
- Formulation of the electronic problem in a tight binding representation using semi-empirical interatomic/interionic matrix elements. For details the reader is referred to W. A. Harrison's account of solid state matrix elements [20]. In many applications, semi-empirical parametrizations of matrix elements are combined with ZDO-type calculations. Though such

schemes are in the spirit of self-consistent field HF calculations, they may include correlation contributions due to the fitting of matrix elements either to empirical expressions or to elaborate *ab initio* calculations. However, this inclusion of correlation is hardly under control.

- Use of pseudopotentials to represent the chemically inert core electrons of an atom. The true valence orbitals may be replaced by nodeless pseudoorbitals varying smoothly in the core region which can effectively reduce the necessary number of basis functions. Both types of orbital agree in the valence region. The pseudopotentials are constructed by requiring that the orbital energies of pseudo- and true valence orbitals are identical. The practical application of pseudopotentials depends on their transferability from a single atom to an atom being part of a molecule or solid. Essentially the transferability corresponds to the frozen core approximation. In particular the *ab initio* angular-dependent atomic pseudopotentials (or effective core potentials, ECP) of Hay and Wadt [21] and Stevens, Basch and Krauss [22] have been proved to be useful in molecular *ab initio* calculations. They can also be used in embedded cluster calculations, which are designed to simulate local electronic defect properties (see Sect. 2.1.2).
- Use of Slater's Xα-technique [18] by means of which the non-local exchange operator is approximated by the local substitute

$$V_{x\alpha}(\boldsymbol{r}) = -6\alpha \left(\frac{3}{4\pi} \varrho(\boldsymbol{r}) \right) . \tag{2.12}$$

  From a modern point of view this is a local density approximation to the exact HF exchange operator. In particular the orbital dependence in (2.7) has been averaged such that each electron experiences this exchange potential.

Four approaches are possible in order to perform rigorous calculations including electron correlations [17, 16, 23, 24, 25]. With either method only a part of the total correlation effects can be included in practice:

- In the configuration interaction (CI) description the total N-electron wavefunction is expanded with a suitable number of Slater determinants. The method is based on the observation that the set of possible N-electron Slater determinants generated from an orthonormal one-electron basis set forms an orthonormal basis of the antisymmetric N-electron Hilbert space. In practice the Hartree–Fock orbitals are conveniently chosen as such a one-electron basis. The Slater determinants occurring in the expansion of the exact N-electron wavefunction may be arranged according to the number of excited electrons with respect to the chosen Hartree–Fock *reference* state function. Calculations restricted to include single and double electron excitations are referred to as SDCI, for instance. Multi-reference CI calculations employ more than one reference state. In practice, all CI expansions must be truncated after a comparatively short number of terms, leading to the well-known size consistency problems of CI calculations (see

[17], for example). There are prescriptions to achieve approximate or exact size consistency [17], e.g. use of the Davidson correction formula and of (coupled) pair theories, respectively. Finally, the multiconfiguration self-consistent field (MCSCF) method and the generalized valence bond (GVB) descriptions are also closely related to the CI approach.

- Application of Rayleigh Schrödinger perturbation theory to a many-body (N-electron) Hamiltonian. The method is often referred to as Møller–Plesset perturbation theory (MP). In this approach the Hartree–Fock Hamiltonian consisting of the sum of the one-electron Fock operators is chosen as the zeroth-order Hamiltonian. Using a diagrammatic representation of the perturbation expressions J. Goldstone proved the *linked cluster theorem*. The theorem states that the perturbation expansion of the total energy can be represented solely by the linked diagrams. Since these diagrams are proportional to the number of electrons in the system, the MP perturbation theory is size-consistent in any order.

The required computing resources allow us to apply the above two approaches only to small systems of molecular dimensions.

- The analysis of the one-particle many-body Green's function allows the determination of the ground state energy of N-electron sytems and of elementary excitation energies. Quasi-particle concepts are related to this Green's function approach. The major difficulties of this technique refer to the determination of the self-energy operator Σ. In Hedin's GW approximation [26], which includes the screened Coulomb interaction W between two electrons to first order, the self-energy is treated as a functional of the one-electron Green's function G, $\Sigma = \Sigma[G]$.
- Application of density functional theory (DFT), a sketch of which is given below. DFT is by construction a theory of the electronic ground state of a system. In principle, DFT is exact, but simplifying assumptions are inevitable in all practical calculations. Besides the inclusion of electron correlation the most powerful advantage of DFT is the significant reduction of necessary computer power. This is related to the formulation of the problem in terms of the electron density. There is no general prescription for the application of DFT to excited states. Jones and Gunnarsson [24] have given an extensive review of recent successful applications of DFT to molecular and solid state systems.

DFT [27] is based on a variational approach using the electron density $\varrho(\boldsymbol{r})$ (instead of wavefunctions) as the fundamental variable:

$$E[\varrho] = \int \varrho(\boldsymbol{r}) V_{\mathrm{ext}}(\boldsymbol{r}) \mathrm{d}^3 r + F[\varrho] \,. \tag{2.13}$$

$F[\varrho]$ is a universal functional (i.e. independent of the external field V_{ext} of the nuclei) representing all kinetic energy and electron–electron interaction effects. Following Levy [28] it is defined by the expression:

$$F[\varrho] = \min_{\Psi\{\varrho\}} \langle \Psi\{\varrho\} | T + V_{\text{ee}} | \Psi\{\varrho\} \rangle \,. \tag{2.14}$$

The domain of F is given by the N-representable densities [28], i.e. densities ϱ which can be generated from N-electron wavefunctions Ψ by integration. In (2.14) wavefunctions leading to ϱ are denoted by $\Psi\{\varrho\}$. Minimization of $E[\varrho]$ with respect to the (N-representable) electron density ϱ yields the ground state energy [24, 28]:

$$E_g = \min_{\varrho} E[\varrho] = E[\varrho_g] \,. \tag{2.15}$$

Introducing a fictitious model system of non-interacting electrons allows us to formulate the DFT variational problem in terms of effective one-particle equations (Kohn–Sham (KS) equations) analogous to Hartree–Fock theory [29, 30]. The universal functional $F[\varrho]$ may be rewritten as:

$$F[\varrho] = T_{\text{s}}[\varrho] + \frac{1}{2} \int \int \frac{\varrho(\boldsymbol{r})\varrho(\boldsymbol{r}')}{|\boldsymbol{r} - \boldsymbol{r}'|} \mathrm{d}^3 r \, \mathrm{d}^3 r' + E_{\text{xc}}[\varrho] \,. \tag{2.16}$$

$T_{\text{s}}[\varrho]$ denotes the kinetic energy functional for a system of non-interacting electrons and the defined energy functional $E_{\text{xc}}[\varrho]$ is called the exchange correlation functional. Variation of the total energy yields the required Kohn–Sham equations:

$$\left(-\frac{1}{2}\Delta - V_{\text{ext}}(\boldsymbol{r}) + \int \frac{\varrho(\boldsymbol{r}')}{|\boldsymbol{r}' - \boldsymbol{r}|} \mathrm{d}^3 r' + \frac{\delta E_{\text{xc}}[\varrho]}{\delta \varrho(\boldsymbol{r})} \right) \psi_k(\boldsymbol{r}) = \varepsilon_k \psi_k(\boldsymbol{r}), \tag{2.17}$$

$$\left(-\frac{1}{2}\Delta + \mathcal{V}_{\text{KS}}[\varrho](\boldsymbol{r}) \right) \psi_k(\boldsymbol{r}) = \varepsilon_k \psi_k(\boldsymbol{r}) \,, \tag{2.18}$$

$$\varrho(\boldsymbol{r}) = \sum_{n}^{\text{occupied}} |\psi_n(\boldsymbol{r})|^2 \,. \tag{2.19}$$

The remarks on basis functions which have been quoted in the context of HF theory also apply to the KS equations. Notably, the number of required two-electron matrix elements reduces to L^3 in comparison to HF theory.

$\varrho(\boldsymbol{r})$ as evaluated on the basis of (2.19) minimizes the energy functional according to (2.15) [1]. Differently from HF, the one-electron energies and orbitals do not have any physical significance, since they are part of a mathematical construction. In the exact DFT only the energy of the highest occupied orbital attains a physical interpretation corresponding to the negative (minimal) ionization energy [30]. The Kohn–Sham reformulation of the problem is exact: all interactions between the electrons of the real system are hidden in

[1] This zero-temperature scheme can be successfully applied to most of the non-metallic systems. In principle, however, the generalization to finite temperatures could be achieved by weighting the one-electron densities in (2.19) according to a Fermi distribution function and extending the sum to include also virtual one-electron orbitals [30]. In practice, the sum should remain finite.

the unknown exchange correlation potential $\delta E_{\mathrm{xc}}[\varrho]/\delta\varrho(\boldsymbol{r})$. In practice the KS equations can be solved if the local density approximation (LDA, or LSDA in the cases where spin polarization is considered) is employed to model the exchange correlation functional:

$$E_{\mathrm{xc}}^{\mathrm{LDA}}[\varrho] = \int \varrho(\boldsymbol{r})\varepsilon_{\mathrm{xc}}^{\mathrm{LDA}}(\varrho(\boldsymbol{r}))\mathrm{d}^3 r\,. \tag{2.20}$$

In this approximation the exchange correlation energy density $\varepsilon_{\mathrm{xc}}$ (and also the exchange correlation potential) depends locally on the charge density ϱ. Slater's Xα potential (see (2.12)) fits into this scheme; approaches including correlation effects were suggested by Hedin and Lundqvist [31], von Barth and Hedin [32] and Ceperly and Alder [33], for instance. The exchange correlation energy density can in principle be written in terms of an exchange correlation charge density:

$$\varepsilon_{\mathrm{xc}}[\varrho](\boldsymbol{r}) = \frac{1}{2}\int \frac{\varrho_{\mathrm{xc}}(\boldsymbol{r},\boldsymbol{r}')}{|\boldsymbol{r}-\boldsymbol{r}'|}\mathrm{d}^3 r'\,. \tag{2.21}$$

Though ϱ_{xc} is normalized within the LD type approximations (compare with (2.10)), it does not obey the required asymptotic behaviour $\propto \frac{1}{r}$. Instead it falls off exponentially, reflecting the asymptotic behaviour of the one-electron orbitals [30] leading to spurious self-interaction contributions.

There are several approaches to improve on the LD type approximations (see [16, 30], for instance). Only one of these procedures is quoted in this context: Becke [34] proposed an exchange functional which incorporates the correct asymptotic behaviour of the exchange density. Using one adjustable parameter the exchange functional has been fitted to reproduce the exchange energies of six noble gas atoms. Lee, Yang and Parr [35] turned the Colle–Salvetti correlation energy formula [36] for the two-electron helium atom into an explicit functional of ϱ. The combined "BLYP" exchange correlation functional has been found very useful in atomic or molecular DFT calculations. This functional, which corresponds to the generalized gradient approximation (GGA) level of accuracy (see e.g. [37]), can also be employed within embedded cluster calculations.

Density functional theory is by construction a theory of electronic ground states. In spite of this feature the exact ground state density ϱ_g also includes information on all excited states. The qualitative argument is as follows: the cusps of ϱ_g determine the nuclear positions; $|\nabla\varrho_g|$ taken at these positions should give the nuclear charges. Thus, in the absence of applied external fields, the total Hamiltonian is known and the diagonalization of which yields the excited states. Consequently, the excited states and their energies may be considered as functionals of ϱ_g. However, there are no straightforward and easy-to-handle procedures to turn this information into operational prescriptions. For a detailed discussion of related questions see [30, 38]. Note that even the ground state energy functional may carry some information on excited states: according to Perdew and Levy [39], every extremum density of the ground state energy functional yields the exact energy of a stationary state. Whereas the absolute minimum corresponds to the ground state, the other extrema represent a subset of the excited states. Exact densities which do not extremize the ground state functional, however, provide lower bounds to the corresponding excited state energies, i.e. $E[\varrho_i] < E_i$, where i refers to any such excited state. Further, the set of extremum densities of the ground state functional generally forms a subset of all stationary densities which may be obtained by applying the usual Kohn–Sham approach. However, if the Kohn–Sham densities provide reasonable approximations to exact densities one may use the above-stated inequality to estimate the requested energy separations.

A very recent elaboration of ordinary density functional theory refers to the density polarization functional theory (DPFT) [40]. It applies to the calculation of ground state properties of a periodic insulating solid which is exposed to a perturbationally small and homogeneous electric field. It turns out, that the perturbing potential is not a unique functional of the (periodic) density change, but also of the change of the macroscopic polarization. The proof of this important result is based on the Hylleraas minimum principle [41], which is related to second-order perturbation theory. It is emphasized that the original Hohenberg–Kohn theorem [27] is not applicable in this situation due to the impossibility of a ground state in the presence of an electric field, which renders invalid the original proof of DFT: a translation of the crystal against the field by multiples of the lattice constant always lowers the energy. It was shown that the application of perturbation theory and the choice of a sinusoidal electrostatic potential

$$V(\boldsymbol{r}) = \lim_{q\to 0} \boldsymbol{E}\cdot\boldsymbol{r}\frac{\sin(q\cdot r)}{q\cdot r} \tag{2.22}$$

can bypass this problem.

As a consequence of these results the exchange correlation functional depends on the density and on the polarization. This leads to an $\mathcal{O}(1/q^2)$-dependence of the exchange correlation kernel for $q \to 0$ [42], which is not observed in all known LDA- and GGA-type XC functionals. This divergence, however, is needed for correct computations of dielectric material constants. It may be approximated by applying LDA combined with a constant energy gap correction (scissor correction) [43].

The remaining subsections consider the application of electronic structure calculations to perfect and defective crystals.

2.1.1 Perfect Crystals

Perfect crystals exhibit full translational symmetry. The major interest is in the electronic ground state properties of these systems.

The present state-of-the-art corresponds to self-consistent DFT-LDA calculations, which are more convenient than HF-based investigations (see above). $\boldsymbol{k}$-space methods are used to find accurate one-particle Bloch orbitals which are solutions of the appropriate Kohn–Sham equations. Commonly, the Muffin Tin approximation (MT) is used to describe the crystal potential: within a sphere surrounding each nucleus the potential is taken spherically symmetrical and constant in the remaining interstitial region, i.e. outside the atomic spheres. Obvious shortcomings are related to the neglected non-spherical contributions at the boundary of each MT atomic sphere and to non-constancy in the interstitial region. For densely packed materials the accuracy of the MT approximation is remarkable; it is exemplified by the reproduction of experimental Fermi surfaces to within 0.01 Ry in suitable cases [44]. The introduction of basis functions leads to a matrix equation which is non-linear in the one-electron energies ε_i:

$$\sum_l (H_{kl}(\varepsilon_i) - \varepsilon_i S_{kl}(\varepsilon_i))C_{il} = 0\,. \tag{2.23}$$

The non-linearity arises since the partial wave solutions (for arbitrary ε-values) within each MT sphere α,

$$\varphi_{\alpha,lm}(\boldsymbol{r} - \boldsymbol{R}_\alpha; \varepsilon) = R_{\alpha,lm}(|\boldsymbol{r} - \boldsymbol{R}_\alpha|; \varepsilon) Y_{lm}(\Omega)\,, \tag{2.24}$$

are used to form Bloch-type basis functions with the MT parts matching at least continuously onto the respective interstitial functions. Many state-of-the-art perfect crystal electronic structure calculations employ one of the following methods:

- Linearized augmented plane wave method (LAPW). This method augments plane waves in the interstitial region with the partial wave solutions of the MT spheres. The basis set indices are given by reciprocal lattice vectors $\boldsymbol{G}$.
- Linearized Muffin Tin orbital method (LMTO) which is related to the earlier KKR Green's function procedure. The interstitial parts of the basis functions correspond to zero kinetic energy ($\kappa \to 0$ instead of $\kappa = \sqrt{\varepsilon - V_{\mathrm{MT}}^{\mathrm{int}}}$) and are thus derived from the Laplace equation. The lm quantum numbers of the MT sphere solutions provide the basis set indices. LMTO strongly resembles LCAO-type procedures.

In both methods the use of energy-independent basis functions allows us to formulate a linear matrix eigenvalue equation (for details see [44]). The original energy-dependent radial basis functions $\varphi(\varepsilon, \boldsymbol{r})$ of APW and KKR, respectively, are replaced by linear combinations of $\varphi(\varepsilon_\nu, \boldsymbol{r})$ and $\partial_\varepsilon \varphi(\varepsilon, \boldsymbol{r})|_{\varepsilon=\varepsilon_\nu}$ for appropriately specified and fixed energy parameters ε_ν. Further, LMTO is often combined with the atomic sphere approximation (ASA) by means of which the interstitial region is totally neglected through the use of overlapping MT-type atomic spheres. Because of the simplifications related to the interstitial region, LMTO calculations are even more restricted to densely packed materials than LAPW-based investigations. In order to accurately model the ferroelectricity of ABO_3 perovskites, full-potential methods, such as FLAPW (e.g. [45]) or FLMTO [46], must be employed. Full-potential calculations invoke the exact crystal potential replacing the MT potential; in particular, full potentials are space-filling and anisotropic. Investigations on $KNbO_3$ and $KTaO_3$ [47] have shown that the ASA approximation is not sufficiently accurate to model the small energy differences related to ferroelectric distortions. However, calculated LMTO-ASA and FLMTO band structures seem to be in acceptable qualitative agreement [47, 48].

Another description of the electronic structure is provided by the pseudopotential approach. By construction, a pseudopotential is the sum of the effective one-electron potential as appearing in the KS equations and of a term representing the orthogonality between core and valence states. The pseudopotential gives rise to smooth valence pseudoorbitals. A pure plane wave basis may conveniently be used to expand the valence pseudoorbitals. By generalizing the norm-conservation conditions for pseudoorbitals one arrives at the ultra-soft pseudopotential schemes (see [49], for example), which allow us to use a particularly small number of plane waves. Regarding the modelling of FE-PT, pseudopotential-based calculations turned out to be of comparable quality to FLAPW investigations [49].

Finally, one can perform LCAO-type (linear combination of atomic orbitals) or LCGTO-type (linear combination of Gaussian-type orbitals) band structure calculations. Since these calculations do not involve any restricting assumptions concerning the effective one-electron potential, they belong to full-potential treatments.

In spite of lacking a rigorous justification, the Kohn–Sham eigenvalues $\varepsilon(\boldsymbol{k})$ obtained from band structure calculations are interpreted as the proper crystalline energy bands. The extent to which this interpretation is reasonable must be inferred from more sophisticated many-particle theories [16, 24]. At least the Fermi energy is exact in DFT, since it can be written as the difference of ground state energies. The band gap can be calculated exactly within DFT, if it is defined in terms of the exact ground state energy $E_0(N)$ of an N-electron system:

$$E_{\mathrm{g}} = E_0(N+1) + E_0(N-1) - 2E_0(N)\,. \tag{2.25}$$

However, in all practical investigations the band gap is determined as the Kohn–Sham orbital energy difference of the lowest unoccupied and the highest occupied one-electron levels. Due to the discontinuity of the exchange correlation potential this method can lead to substantial underestimations of band gap energies (up to 50%). This principal deficiency has been further demonstrated on the basis of a discrete lattice DFT employing a Hubbard-type Hamiltonian [50]. The deviations between the Kohn–Sham band gap and the exact gap increase abruptly with increasing Hubbard-U parameter, measuring the electron correlations. For further aspects of the gap problem, see the review article of Jones and Gunnarsson [24]. Very recently, the band gap and optical transitions related to MgO were successfully simulated by application of the GW approximation [51].

Band structure calculations can be used to determine linear (e.g. [52]) and, in principle, also non-linear optical material coefficients [53]. Since these properties are mainly based on electronic excitations, it is important to correct the LDA eigenenergies of excited one-electron states. This can be accomplished either by simply shifting excited energy states (scissor operator technique) or by applying more sophisticated corrections due to advanced quasi-particle concepts (e.g. the *GW* formalism [16]). The investigations of Ching et al. [52] suggest that a rigid energy shift can be too crude: in $LiNbO_3$ the changes in single-electron states are both energy- and $\boldsymbol{k}$-dependent.

2.1.2 Defective Crystals

We restrict our considerations to point defects. Particular interest is devoted to the electronic aspects of the formation of such defects and to possible energy levels in the band gap. Defects destroy the translational symmetry of perfect crystals. $\boldsymbol{k}$-space methods are therefore inappropriate in order to treat isolated point defects. They are applicable only in those cases where a periodic array of defects has been imposed, thereby reinstalling translational symmetry.

Basically, three different theoretical approaches to defects can be distinguished: the supercell method, Green's function techniques and embedded cluster calculations. The latter two methods are closely connected to each other. This will be seen subsequently.

Embedded cluster calculations allow the study of the local electronic defect structure of charged point defects, taking lattice relaxations fully into account. This can be accomplished by representing the embedding lattice on the basis of a pair potential shell model description. Defect-induced lattice distortions are particularly important in ionic or semi-ionic materials, such as the oxides discussed in this book. Green's function approaches and supercell calculations, on the other hand, are much more restricted to neutral defects and are less well suited to including defect-induced lattice perturbations. Their success is particularly related to a description of defects in semiconductors.

All methods have their own relative merits. It is thus recommended to consider all approaches to aim at a complete picture of the electronic properties of defects.

Supercell Method. This approach extends the perfect crystal calculations to defective systems. Supercells containing several unit cells of the investigated material and a specified defect are periodically arranged to form a supercrystal (Fig. 2.1).

As in perfect crystals, $\boldsymbol{k}$-space techniques based on HF or DFT effective one-particle descriptions are applicable. Because of the periodic boundary conditions there are no artificial surface effects in this approach. However, large supercells are necessary in order to avoid direct and indirect defect–defect interactions. The latter are mediated by the crystal lattice (e.g. lattice

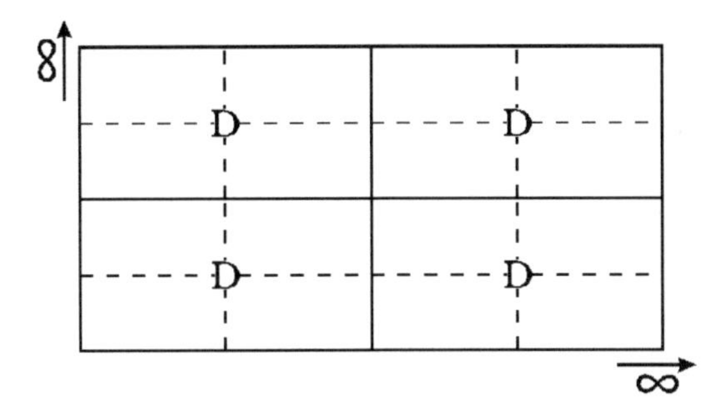

Fig. 2.1. Periodic array of supercells simulating a defective crystal

deformation fields). In practice, lattice relaxations can only be approximately taken into account, e.g. by including nearest neighbour relaxations. However, the lattice deformation fields are frequently of a longer range.

Crystal stability requires the use of neutral supercells. As a consequence, the defect considered must be neutral or one must introduce appropriate local charge-compensating defects. For example, one could study the local complex $Fe^{3+}_{Nb} - V_O$ in $KNbO_3$, but not isolated Fe^{3+}_{Nb}. The definition of neutralizing homogeneous background charges would remedy this disadvantage; however, in many situations such an artificial compensator is not very satisfactory.

The required neutrality condition is not sufficient to obtain reliable results. The success of any supercell calculation depends to a considerable extent on the convergence properties of the infinite electrostatic sums which are determined by the lowest order non-vanishing multipole moments per cell. Corresponding questions have been recently investigated by Makov and Payne [54]. In the present context it suffices to remark on calculations involving supercells with non-vanishing dipole moments. The above-mentioned defect calculations belong in this category, as do the numerous simulations of ferroelectric phases of oxide perovskites (see Sect. 3.1). Since only the second derivative of the electrostatic potential converges absolutely, these calculations are affected by the indeterminacy of an unknown constant electric field. Its magnitude depends on the definition of the supercell. Useful results can be obtained only if the *zero field approximation* [54] is applied, for which the physical justification may be given by the assumption of surface-adsorbed impurities equilibrating the crystalline electrostatic potential.

Supercell calculations are particularly useful to study the effects of high defect concentrations. However, the periodicity of defects seems to be artificial when compared with realistic situations with random defect distributions. KLT and KTN have recently been investigated using a supercell approach and DFT-LDA [47, 55].

Green's Function Description. In principle all approaches based on Green's functions are exact, since they provide the correct coupling between the defect and the bulk crystal. The methods allow the study of bound defect states as well as resonances. Bound states are characterized by energy levels within the gap, whereas resonances occur within the allowed energy bands.

By definition a Green's function corresponds to the resolvent operator of a one-electron Hamiltonian. Two different principal descriptions can be employed, i.e. the *perturbed crystal* and the *perturbed cluster* treatments (see [56] for a detailed review).

1. *Perturbed crystal approach:* The perturbed crystal approach considers the defect-induced potential as a perturbation of the host crystal. One starts with the one-electron Hamiltonian of the defective system

$$H = H_0 + U\,, \tag{2.26}$$

where H_0 is the perfect crystal Hamiltonian and U the perturbing potential due to a single point defect:

$$U = \mathcal{V}(\boldsymbol{r}, \varrho(\boldsymbol{r})) - \mathcal{V}(\boldsymbol{r}, \varrho_0(\boldsymbol{r}))\,. \tag{2.27}$$

$\mathcal{V}$ denotes the effective potential appearing in the one-electron HF- or KS equations. The relevant Green's functions correspond to the resolvent operators of the perfect and defective one-electron Hamiltonians:

$$G_0(\varepsilon) = \lim_{\eta \to 0^+} (\varepsilon + i\eta - H_0)^{-1} \tag{2.28}$$

$$G(\varepsilon) = \lim_{\eta \to 0^+} (\varepsilon + i\eta - H)^{-1} \tag{2.29}$$

The perfect lattice Green's function G_0 and its defective counterpart are related by the Dyson equation which is central to the perturbed crystal approach:

$$G(\varepsilon) = G_0(\varepsilon) + G_0(\varepsilon)UG(\varepsilon) = (1 - G_0(\varepsilon)U)^{-1}G_0(\varepsilon)\,. \tag{2.30}$$

According to (2.30) $G(\varepsilon)$ becomes singular for all non-perturbed bulk states and for defect-induced states. Defect levels are obtained by analyzing the condition

$$D(\varepsilon) = \det(1 - G_0(\varepsilon)U) = 0\,. \tag{2.31}$$

Important information concerning defect levels is provided by the density of states (DOS) $N(\varepsilon)$:

$$N(\varepsilon) = -\frac{1}{\pi}\mathrm{Tr}(\mathrm{Im}\; G(\varepsilon))\,. \tag{2.32}$$

Energy integration,

$$q = e\int_{-\infty}^{\varepsilon_F} N(\varepsilon)\mathrm{d}\varepsilon\,, \tag{2.33}$$

yields the total electron charge, so the charge density operator is therefore given by

$$\varrho = e \sum_{i}^{\text{occ}} |\psi_i\rangle\langle\psi_i| = -\frac{e}{\pi} \int_{-\infty}^{\varepsilon_{\mathrm{F}}} \mathrm{Im}\ G(\varepsilon)\mathrm{d}\varepsilon\,. \tag{2.34}$$

The charge density matrix follows as:

$$\varrho(\boldsymbol{x}, \boldsymbol{y}) = e \sum_{i}^{\text{occ}} \langle \boldsymbol{x}|\psi_i\rangle\langle\psi_i|\boldsymbol{y}\rangle = -\frac{e}{\pi} \int_{-\infty}^{\varepsilon_{\mathrm{F}}} \mathrm{Im}\ G(\boldsymbol{x}, \boldsymbol{y}; \varepsilon)\mathrm{d}\varepsilon\,. \tag{2.35}$$

In order to calculate $D(\varepsilon)$ the operators G_0 and U must be expanded within a suitable system of basis functions. The Green's function method is extremely useful in all cases where the perturbing potential U may be assumed to be well localized, leading to manageable matrix equations of sufficiently small dimensions. Due to effective screening effects the localization condition is fulfilled in metals and semiconductors. The situation is different in (semi-)ionic systems where defect-induced potentials with significant Coulomb contributions are of long-range nature. Moreover, the inclusion of lattice relaxation terms should lead to more extended defect-induced potentials. Lattice relaxation may be important for the stability of certain charge states of defects, since it shifts the defect energy levels with respect to the Fermi level of the crystal.

In principle, (2.27) should be used to self-consistently determine the *ab initio* perturbing potential U from the charge density ϱ of the defective lattice; ϱ_0 denotes the corresponding and known density of the perfect crystal. To my knowledge there are no *ab initio* investigations on defects in complex oxide materials based on Green's functions. Instead, most approaches like the studies of Selme et al. [57, 58, 59] as well as Fisenko and Prosandeyev [60, 61, 62] in ABO_3 perovskites employ short-range *ad hoc* defect potentials which are independent of ϱ. In both cases the set of basis functions was chosen to be of LCAO-type (tight binding model) with matrix elements taken from the earlier work of Mattheiss [63]. Further approximations concern the restriction to first-neighbour interactions and to the inclusion of only oxygen 2p- and B-cationic d-states. Local lattice relaxations may be taken into account within these tight binding approaches on the basis of distance variations of interatomic solid state matrix elements, as proposed by Harrison [20]. Extended lattice relaxation effects can only be accounted for on the basis of simplified screening arguments.

2. *Perturbed cluster approach:* In this description the crystal is divided into a cluster and the embedding environment. The influence of the embedding crystal regions on the electronic structure of the cluster is at the heart of this method. The perturbed cluster approach represents the natural link between embedded cluster calculations (Sect. 2.1.2) and Green's function treatments.

The perturbed cluster considerations due to Löwdin [64] start with a set of localized basis functions. In the following the indices c and e refer to the cluster and environment, respectively. After introducing the abbreviating notation $Q = \varepsilon I - H$ (with I being the identity matrix) the one-electron Schrödinger type equation of the total crystal reads as:

$$\begin{pmatrix} Q_{\mathrm{cc}} & Q_{\mathrm{ce}} \\ Q_{\mathrm{ec}} & Q_{\mathrm{ee}} \end{pmatrix} \begin{pmatrix} \psi_{\mathrm{c}} \\ \psi_{\mathrm{e}} \end{pmatrix} = 0 . \tag{2.36}$$

Inserting the expression for ψ_{e} resulting from the second-row equation of (2.36) into the first-row equation yields a Schrödinger-type equation for the cluster-projected wavefunction ψ_{c}:

$$(Q_{\mathrm{cc}} - Q_{\mathrm{ce}}(Q_{\mathrm{ee}})^{-1} Q_{\mathrm{ec}})\psi_{\mathrm{c}} = 0 . \tag{2.37}$$

This perturbed cluster equation is exact and may be applied to all types of solids. It determines the cluster-projected part of the one-electron eigenstates corresponding to the exact eigenenergies ε. The cluster-projected Hamiltonian Q_{cc} does not describe an isolated cluster, but rather a cluster embedded within the surrounding crystal region. Corresponding aspects will be further discussed in Sect. 2.1.2. In addition the second term in (2.37) represents an effective, energy-dependent potential which models an electron transfer between the cluster and the environment:

$$V_{\mathrm{cc}}^{\mathrm{t}} = Q_{\mathrm{ce}}(Q_{\mathrm{ee}})^{-1} Q_{\mathrm{ec}} . \tag{2.38}$$

Q_{ec} and Q_{ce} define the hopping matrix elements for an electron leaving and re-entering the cluster, respectively; $(Q_{\mathrm{ee}})^{-1}$ is the propagator modelling the electron motion through the embedding region. $V_{\mathrm{cc}}^{\mathrm{t}}$ reduces to a cluster surface operator if the hopping matrix elements are of short range relative to the cluster dimensions. Due to the electron mobility $V_{\mathrm{cc}}^{\mathrm{t}}$ is indispensable in metals. However, in insulating materials including oxides this operator may safely be neglected (Sect. 2.1.2).

For applications of this perturbed cluster method simplifying approximations need to be introduced in order to treat $V_{\mathrm{cc}}^{\mathrm{t}}$. Most importantly one considers V^{t} to be unaffected by the presence of a defect, i.e. $V_{\mathrm{cc}}^{\mathrm{t}} = V_{\mathrm{cc}}^{\mathrm{t}\,0}$. This relation immediately leads to the Baraff–Schlüter formula [65]:

$$V_{\mathrm{cc}}^{\mathrm{t}} = Q_{\mathrm{cc}}^{0} - (G_{\mathrm{cc}}^{0})^{-1} . \tag{2.39}$$

According to (2.39) only perfect crystal quantities enter the calculation of $V_{\mathrm{cc}}^{\mathrm{t}}$; in particular the cluster-projected perfect lattice Green's function can be taken from band structure calculations. Inserting (2.39) into (2.37) recovers the perturbed crystal relation:

$$\psi_{\mathrm{c}} = G_{\mathrm{cc}}^{0}(Q_{\mathrm{cc}}^{0} - Q_{\mathrm{cc}})\psi_{\mathrm{c}} = G_{\mathrm{cc}}^{0} U_{\mathrm{cc}} \psi_{\mathrm{c}} , \tag{2.40}$$

where U_{cc} is the defect-induced potential (see (2.27)), if the cluster extension corresponds to the range of U.

A completely different type of perturbed cluster calculations is defined by the corrective operator methods. These descriptions cannot be linked to the perturbed crystal approach, which refers to scattering theory. Multiplicative and additive corrective operator treatments are known [56]. These methods are based on relations between the cluster-projected Green's function G_{cc} and the finite-cluster quantity $(Q_{cc})^{-1}$ which are established by the respective corrective operators. A brief discussion is now given for the additive case, which forms the basis of the perturbed cluster approach of Pisani et al. [66, 67]. For a recent application of this method to carbon impurities in silicon, see [68].

One starts from the equation $QG = I$, which yields:

$$Q_{cc}G_{cc} + Q_{ce}G_{ec} = I_{cc} \tag{2.41}$$

$$Q_{cc}G_{ce} + Q_{ce}G_{ee} = 0 . \tag{2.42}$$

Inserting

$$G_{ec} = -G_{ee}Q_{ec}(Q_{cc})^{-1} , \tag{2.43}$$

which follows from (2.42) upon using $G_{ce} = (G_{ec})^{\mathrm{T}}$, into (2.41) gives:

$$\begin{aligned} G_{cc} &= (Q_{cc})^{-1} + (Q_{cc})^{-1}Q_{ce}G_{ee}Q_{ec}(Q_{cc})^{-1} \\ &= (Q_{cc})^{-1} + \Delta G_{cc} . \end{aligned} \tag{2.44}$$

Equations (2.32) and (2.35) may be applied to calculate the density of states and the electron charge density, respectively. The corresponding additive corrections involving ΔG_{cc} describe the effects induced by electron transfer effects.

ΔG_{cc} in (2.44) contains the propagator G_{ee} which is related to the embedding medium. In order to obtain applicable prescriptions one must impose the approximation $G_{ee} = G^0_{ee}$ [67, 56]. Besides neglecting defect-induced perturbations of the embedding crystal this approach is affected by a further more significant disadvantage: approximations to ΔG_{cc} are not able to subtract the poles of the finite-cluster Green's function $(Q_{cc})^{-1}$ and to add the exact poles of the total defective crystal [56]. It is noted in this context that the energy eigenvalues of a finite cluster can deviate substantially from their proper crystalline counterparts (see also Sect. 2.1.2). This is particularly true if the finite clusters contain transition metal impurities.

Embedded Cluster Calculations. By definition, embedded cluster calculations omit the inclusion of the electron transfer potential given in (2.38). Certainly, this potential plays a vital role in metallic systems, but it is assumed to be negligible in insulating materials, which also include oxides.

Therefore embedded cluster calculations are based on a real-space description of a small crystalline fragment; commonly outer crystal spheres embedding the cluster are treated more approximately. Quantum chemical approaches are applied to simulate the defect and a properly chosen number of neighbouring ligand ions forming the cluster. *Ab initio* MO calculations employ Hartree–Fock theory (which may be extended to include electron correlations), but density functional descriptions are also possible.

Embedded cluster treatments may formally be justified on the basis of localized one-electron Wannier-type orbitals replacing the canonical delocalized Bloch eigenstates of the crystal. For non-defective insulating crystals there exist unitary transformations between the occupied Bloch orbitals and a corresponding set of localized orbitals [69]. In this case the total electron density $\varrho(\boldsymbol{r})$ can be written as

$$\frac{1}{2}\varrho(\boldsymbol{r}) = \sum_{n,\boldsymbol{k}}^{\text{occ}} |\psi_{n,\boldsymbol{k}}(\boldsymbol{r})|^2 = \sum_{n,j}^{\text{occ}} |\omega_n(\boldsymbol{r} - \boldsymbol{R}_j)|^2 . \tag{2.45}$$

$\psi_{n,\boldsymbol{k}}(\boldsymbol{r})$ denotes Bloch orbitals and $\omega_n(\boldsymbol{r} - \boldsymbol{R}_j)$ localized Wannier orbitals, n is the band index and $\boldsymbol{R}_j$ is a direct lattice vector. Instead of the $\boldsymbol{R}_j$-centred Wannier orbitals one can also introduce an orthonormal basis set consisting of localized Wannier orbitals centred at the nuclear positions $\boldsymbol{r}_\kappa$, see also the footnote [2]. Such orbitals are conveniently introduced using the site symmetry concept (see, for example, [70]). The site symmetry group $\mathcal{G}_{\boldsymbol{r}_\kappa}$ of a particular Wyckoff position $\boldsymbol{r}_\kappa$ in the considered crystal consists of all operations $\mathcal{O} \in \mathcal{G}_0$ (with $\mathcal{G}_0$ being the space group) satisfying $\mathcal{O}\boldsymbol{r}_\kappa = \boldsymbol{r}_\kappa$. $\mathcal{G}_{\boldsymbol{r}_\kappa}$ belongs to one of the known 32 point symmetry groups. In perfect crystals the atoms are located at Wyckoff positions. Atom-centred orbitals may be constructed as basis functions transforming as partners of the irreducible representations of the site symmetry group $\mathcal{G}_{\boldsymbol{r}_\kappa}$. Then (2.45) reads as:

$$\frac{1}{2}\varrho(\boldsymbol{r}) = \sum_{n,\boldsymbol{k}}^{\text{occ}} |\psi_{n,\boldsymbol{k}}(\boldsymbol{r})|^2 = \sum_{l,\kappa}^{\text{occ}} |\varphi_l(\boldsymbol{r} - \boldsymbol{r}_\kappa)|^2 . \tag{2.46}$$

κ labels the atomic sites and l the basis functions of all irreducible representations of $\mathcal{G}_{\boldsymbol{r}_\kappa}$ corresponding to occupied levels.

The use of localized orbitals appeals strongly to chemical intuition about the formation of interatomic bonds. Equation (2.46) allows calculation of the crystalline charge density within the restricted area of a cluster. This representation is exact for crystal-adapted localized orbitals. It is obvious that the localized cluster orbitals are not independent of the crystalline environment.

Conceptually the incorporation of isolated point defects into the cluster does not cause any problems, because the atom-centred spin-orbitals which belong to a point defect replacing a specified crystal atom (or ion) can be chosen orthogonalized with respect to the remaining perfect crystal orbitals. Subsequently, therefore, the clusters may be assumed to be perfect or even defective.

For metals, analogous statements to (2.45) and (2.46) do not hold true, which reflects the importance of the electron transfer potential in these ma-

[2] Here we restrict our considerations to ionic or semi-ionic crystals. In covalent materials, on the other hand, the properly chosen centres of localized orbitals may correspond to sites between pairs of atoms resembling the formation of electronic pair bonds. However, this modification does not change the principal group-theoretical concept.

terials discussed in the previous subsection. Therefore embedded cluster calculations are not applicable to metallic systems.

Considering insulating solids, (2.46) forms the basis to divide the total crystal into the cluster and the embedding environment, with the electron charge density matrix of the crystal given by:

$$\begin{aligned}\varrho(\boldsymbol{x},\boldsymbol{y}) &= \varrho_{\mathrm{c}}(\boldsymbol{x},\boldsymbol{y}) + \varrho_{\mathrm{e}}(\boldsymbol{x},\boldsymbol{y})\varrho(\boldsymbol{x},\boldsymbol{y}) \\ &= \sum_i^{\mathrm{occ}} \varphi_{\mathrm{c},i}(\boldsymbol{x})\varphi^*_{\mathrm{c},i}(\boldsymbol{y}) + \sum_j^{\mathrm{occ}} \varphi_{\mathrm{e},j}(\boldsymbol{x})\varphi^*_{\mathrm{e},j}(\boldsymbol{y})\,. \end{aligned} \tag{2.47}$$

Application of the unitary mapping, which was introduced above and which transforms the occupied Bloch orbitals to localized atomic-centred orbitals, to the canonical one-electron Schrödinger-type HF or KS equations, i.e.

$$H[\varrho]\psi_{n,\boldsymbol{k}} = \varepsilon_{n,\boldsymbol{k}}\psi_{n,\boldsymbol{k}}\,, \tag{2.48}$$

leads to the following set of cluster-related one-electron equations [71, 72]:

$$\begin{aligned}H[\varrho]\varphi_{\mathrm{c},i} = H[\varrho_{\mathrm{c}} + \varrho_{\mathrm{e}}]\varphi_{\mathrm{c},i} &= (H^0_{\mathrm{c}} + V_{\mathrm{M}} + V_{\mathrm{SR}})\varphi_{\mathrm{c},i} \\ &= \sum_j \varepsilon_{ij}\varphi_{\mathrm{c},j}\,. \end{aligned} \tag{2.49}$$

The similarly obtainable equations for the environment-related orbitals are not important for present purposes. In (2.49), H^0_{c} denotes the one-electron Hamiltonian of the isolated cluster, V_{M} the long-range Madelung potential due to environmental point charges and V_{SR} the remaining short-range potential of the embedding medium. V_{SR} comprises higher electrostatic multipole fields as well as repulsive short-range contributions (e.g. ion size effects) acting upon the cluster. The unitary transformation leaves $H[\varrho]$ invariant, since the one-electron Hamiltonian is an operator-valued functional of the density matrix. However, the transformed one-electron equations are no longer diagonal, because the localized orbitals do not represent crystalline eigenstates. For non-defective insulators (2.49) is exact. Though no electron transfer potential enters this equation, there is no contradiction of (2.37), because we are now dealing with truly localized orbitals rather than with the cluster-projected part of the exact crystaline one-electron eigenstates as in (2.37). In (2.37) V^{t} is necessary in order to account for the proper connection of ψ_{c} and ψ_{e}.

Following the localization theory of Adams, Gilbert and Kunz (see [71] and references cited therein) (2.49) can be reformulated as

$$(H^0_{\mathrm{c}} + V_{\mathrm{M}} + V_{\mathrm{SR}})\varphi_{\mathrm{c},i} = \pi_i\varphi_{\mathrm{c},i} + \hat{\varrho}_{\mathrm{c}}W\hat{\varrho}_{\mathrm{c}}\varphi_{\mathrm{c},i}\,, \tag{2.50}$$

where $\hat{\varrho}_{\mathrm{c}} = \sum_j^{\mathrm{occ}} |\varphi_{\mathrm{c},j}\rangle\langle\varphi_{\mathrm{c},j}|$ denotes the cluster-related density operator and W an arbitrary hermitian one-electron operator describing the freedom of choosing the unitary transformation within the set of the occupied cluster orbitals $\{\varphi_{\mathrm{c},i}\}$. The parameters π_i may be interpreted as the one-electron orbital energies of the embedded cluster:

$$(H_c^0 + V_{\text{embed}})\varphi_{c,i} = \pi_i \varphi_{c,i} \,. \tag{2.51}$$

The embedding potential is given by the non-local, configuration- (or energy-) dependent operator:

$$V_{\text{embed}} = V_M + V_{SR} - \hat{\varrho}_c W \hat{\varrho}_c \,. \tag{2.52}$$

The particular choice $W = V_{SR}$ [71] leads to vanishing matrix elements of $V_{SR} - \hat{\varrho}_c V_{SR} \hat{\varrho}_c$ calculated for the occupied cluster orbitals:

$$\langle \varphi_{c,i} | V_{SR} - \hat{\varrho}_c V_{SR} \hat{\varrho}_c | \varphi_{c,j} \rangle = 0 \,, \tag{2.53}$$

with i, j referring to occupied orbitals. Approximating (2.53) as operator identity leads to a minimal embedding, i.e.

$$(H_c^0 + V_M)\varphi_{c,i} = \pi_i \varphi_{c,i} \,. \tag{2.54}$$

This set of equations corresponds to embedding the cluster within a point-charge array. In the case of covalent materials V_M is zero and (2.54) reduces to *in vacuo* cluster calculations.

In the presence of a defect the charge density of the embedding surrounding will deviate from its perfect lattice counterpart. Also, the nuclear positions can be affected by defect-induced changes. Corresponding effects lead to polarizations of the embedding crystal. Due to charged defects, polarizations are particularly important in ionic or semi-ionic materials. Such polarization effects, which give rise to an additional embedding potential term ΔV_{embed} can be conveniently described by employing a shell model-type pair potential representation of the outer crystal regions. Details of this approach are given below.

As an alternative to the embedded cluster treatment presented above, one may employ the theory of electronic separability in order to discuss the embedding of clusters. The most essential idea within this approach, which derives from the group function description of McWeeny [23] and Huzinaga et al. [73], involves separating the total crystal into weakly interacting electron groups, i.e. into the defect cluster and the embedding groups representing ions, atoms or even saturated bonds. To each such group belongs an anti-symmetrized wavefunction which may include correlation contributions. It is obvious that this formulation of embedded clusters starts from localized orbitals, too. The total wavefunction is given by the antisymmetrized product of the building unit wavefunctions:

$$\Phi_{\text{env}} = N_e \, \mathcal{A} \left(\prod_i \Phi_{\text{env},i} \right) , \tag{2.55}$$

$$\Phi_{\text{cryst}} = N_c \, \mathcal{A}(\Psi_{\text{clus}} \Phi_{\text{env}}) \,. \tag{2.56}$$

$N_{e,c}$ are appropriate normalization factors and $\mathcal{A}$ denotes the antisymmetrizer. In this formulation the assumption of weak interactions completely disregards electronic correlations between different groups and also electron transfer effects [23]. The second basic idea of the group function approach involves the strong orthogonality condition, i.e. different group functions are assumed to be orthogonal to each other. This condition follows immediately from (2.55) and (2.56) assuming orthogonalized localized one-electron orbitals. The group function description of crystals results in explicit prescriptions for constructing the embedding potential of a quantum cluster [74, 75]. Kantorovich [74] further derived formulas for the polarization-induced contribution ΔV_{embed}. He proved that corresponding contributions of distant crystal regions comply with classical dielectric continuum theory (Mott–Littleton approach [76]). This result is important, since rigorous applications of the theory to determine ΔV_{embed} are cumbersome. Again, useful approximations to the embedding medium can be introduced by shell model-type representations of the outer crystal regions.

Within both of the discussed embedded cluster formulations the total energy of a crystal can be written as:

$$E_{\text{cryst}} = E_{\text{clus}} + E_{\text{clus-env}} + E_{\text{env}} \,. \tag{2.57}$$

This equation follows either from the group function approach employing the strong orthogonality condition [23, 73] or from the localizing potential treatment of Kunz and Klein [71] assuming orthogonalized one-electron atomic-centred orbitals.

Practical investigations can be classified according to their level of sophistication applied to the cluster embedding:

- *in vacuo* clusters,
- implementation of Watson spheres,
- embedding by point charge arrays,
- embedding by point charge arrays with additional pseudopotentials on boundary ions between the cluster and the environment,
- application of the theory of electronic separability [74, 75].

Generally, *in vacuo* clusters must be very large to reflect any bulk properties. Boundary conditions are introduced to reduce artificial surface effects. In covalent materials it is common practice to attach hydrogen atoms to the cluster surface in order to saturate broken bonds. H atoms are used to separate the surface states from the relevant electronic cluster states. In more ionic crystals, in which we are mainly interested, it is important to include the effects of a Madelung potential. This can be done approximately by implementation of Watson spheres. In many cases only one charged Watson sphere surrounding the whole cluster is used. The many defect studies of Michel-Calendini et al. (e.g. [77]) follow this approach. The introduction of an embedding Watson sphere stabilizes all cluster electrons, but is unable to model Madelung potential differences related to different sites [71]. One should also note that the boundary conditions applied to the tails of the wavefunctions beyond the Watson sphere radius are physically not very satisfactory. This may become particularly crucial for excited states, which are characterized by significant spatial extensions. As a consequence, energy splittings between different states are to a considerable extent dependent on the specific choice of the Watson sphere (see [78] for further discussions). In order to correctly model the Madelung potential, extended point charge arrays embedding the quantum cluster can be employed (other methods such as the Ewald technique are also applicable). This procedure seems to be sufficiently accurate in highly ionic materials with strongly localized electrons, but even in these cases it is advisable to model ion-size effects of the ions located in the immediate neighbourhood of the cluster. This can be accomplished on the basis of suitable pseudopotential methods. The explicit construction of localizing potentials suggested by Kunz and Klein [71, 72] is based on (2.53). However, the use of tabulated effective core potentials (e.g. see [21]) defined on boundary atoms is computationally less costly and facilitates results of remarkable accuracy (see Sect. 3.3.1). For recent applications of the theory of electronic separability to transition metal impurities in alkali halides and alkaline earth fluorides see [79]. Up to now this technique has not been employed to investigate point defects in complex oxides.

For detailed discussions of the embedding problem the reader is referred to [74, 75, 80].

Embedded cluster-type calculations are extremely well suited to investigating the local electronic structure of point defects in ionic or semi-ionic crystals. There is no restriction to neutral defects and it is straightforward to include defect-induced lattice relaxations. This is particularly true if one simulates the cluster environment on the basis of a shell model pair potential representation (see also Sect. 2.2). The computational basis of this approach is given by (2.57), which, for present purposes, may be reformulated as follows:

$$E(\Psi, R_\mathrm{c}, R_\mathrm{e}) = E_\mathrm{SM}^\mathrm{crys}(R_\mathrm{c}, R_\mathrm{e}) + E_\mathrm{QM}^\mathrm{cl}(\Psi, R_\mathrm{c}, R_\mathrm{e}) - E_\mathrm{SM}^\mathrm{cl}(R_\mathrm{c}, R_\mathrm{e}) \,. \quad (2.58)$$

The total energy of the composite system is minimized with respect to the cluster (nuclear) coordinates R_c and to the core and shell coordinates R_e of the outer crystal ions in order to calculate defect energies. In addition, a minimization is needed with regard to the electronic wavefunction Ψ describing the local electronic structure within the cluster region[3]. $E_\mathrm{SM}^\mathrm{crys}(R_\mathrm{c}, R_\mathrm{e})$ is the shell model energy of the total crystal and $E_\mathrm{QM}^\mathrm{cl}(\Psi, R_\mathrm{c}, R_\mathrm{e})$ is the quantum mechanical defect cluster energy including the total Coulomb interaction between cluster species (electrons and nuclei) and the outer crystal ions which are treated within the shell model. To avoid double-counting effects in (2.58) the shell model-type cluster energy $E_\mathrm{SM}^\mathrm{cl}(R_\mathrm{c}, R_\mathrm{e})$ must be subtracted; similar to its quantum mechanical counterpart $E_\mathrm{SM}^\mathrm{cl}(R_\mathrm{c}, R_\mathrm{e})$ includes all Coulomb interactions with the outer crystal ions. Thus, (2.58) mimics the substitution for a defective shell model cluster of its quantum mechanically treated counterpart. Moreover, pairwise short-range potentials are kept to modelling the

[3] The wavefunction must be replaced by the electron density if DFT is employed instead of wavefunction-based methodologies.

corresponding interactions between the cluster and the embedding crystal region.

Figure 2.2 schematically visualizes corresponding computational cycles in order to minimize the energy of the defective crystal. It is common practice to optimize the cluster and embedding lattice configurations on the basis of variable metric methodologies [81].

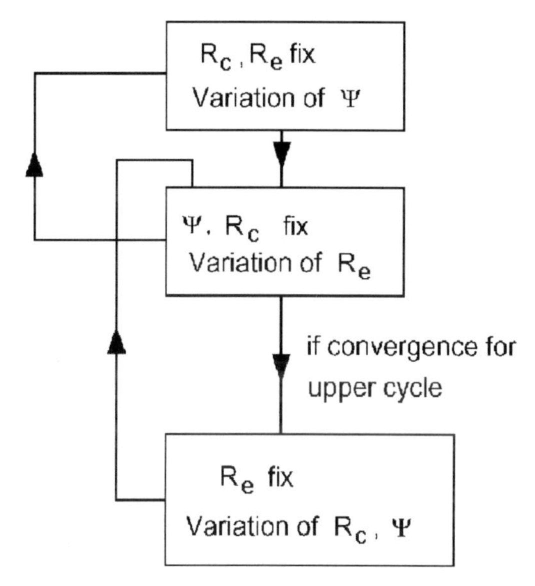

Fig. 2.2. Computational cycles of embedded cluster calculation including lattice relaxation. R_c and R_e denote the cluster and embedding lattice configurations, respectively. Ψ is the electronic state function of the quantum cluster. The embedding lattice is modelled on the basis of a potential representation. Whereas the upper cycle denotes the lattice equilibration step, the lowest box represents the cluster geometry optimization cycle

The complete minimization procedure consists of two cycles: the first cycle performs the geometry optimization of the embedded cluster with all outer ions remaining fixed at their actual positions. For each particular cluster geometry the wavefunction (or electron density) must be updated appropriately. Most of the available *ab initio* molecular codes can be used for this task. However, the programs must be modified as to include the required short-range cluster–lattice interactions. The second cycle in Fig. 2.2 is devoted to the equilibration of the embedding shell model lattice. For corresponding computational details the reader is referred to Sect. 2.2.

Generally the lattice configuration R_e bears a pronounced relation to the cluster wavefunction Ψ. This dependence requires us to employ during the optimization of the lattice configuration R_e the accurate electrostatic potential produced by the quantum cluster. This task may be accomplished by applying a multipole expansion to the electrostatic cluster potential. The required multipoles are derived from the total cluster charge density, which is the output of the appropriate *ab initio* MO calculations. In most situations the embedding lattice relaxation step is based on a shell model description employing point charges to represent the crystal ions (Sect. 2.2). Therefore one naturally introduces point charge simulators [82] of which the positions and charges are adjusted to reproduce the exact cluster multipoles up to a required order. This procedure guarantees the accurate account of the electrostatic cluster potential. More approximate schemes choose the cluster ion charges according to a Mulliken population analysis (MPA, e.g. [17]) or even as formal (integral) charges. The latter two approaches are particularly useful if bare effective core pseudopotentials are used to represent the ions at the cluster boundary. This experience is due to the observations that the pseudopotential ions spatially separate the embedding lattice ions from the explicit cluster charge density and that higher multipole contributions fall off rapidly with increasing distance from the cluster density. A further advantage of using bare effective core potentials at the cluster boundary refers to the fast convergence properties of the total equilibration cycle encompassing both the lattice and the cluster subcycle. In most practical situations it suffices to perform two or three such loops. If not stated otherwise all practical embedded cluster studies reported in the subsequent chapters refer to this minimization procedure.

A slightly different approach to embedded cluster calculations explicitly employs the exact potential energy surface (PES) of the cluster in order to perform cluster and lattice geometry optimization. This method establishes a natural link to general potential simulations described in Sect. 2.2. The technique has been invented for defects with high point symmetry. For details of this method the reader is referred to Sect. 3.3.1, where it has been applied to investigate tetravalent manganese in cubic $BaTiO_3$.

So far, embedded cluster simulations including lattice relaxations have been reported for a number of different defect species in ionic materials, including impurity cations, anion vacancies and hole type defects in alkali halides, elpasolites and in basic binary oxides such as MgO (for example, see [72, 83, 84, 85, 86]). Chapters 3 and 4 present sophisticated embedded cluster calculations for complex perovskite-structured oxides, in which special attention is paid to the effects of electron correlations.

In order to complete the discussion of embedded cluster calculations some remarks on the value of one-electron orbital energies are useful. Orbital energy differences can be highly misleading when calculating energetic separations between ground and excited electronic states, because these energies need not be related to true crystalline one-electron energies. This point is illustrated by tetravalent manganese doped into $BaTiO_3$ on Ti sites, of which further computational details are given in Sect. 3.3.1.

Inspecting the orbital energies of an $(MnO_6)^{-8}$ cluster shows that the singly occupied $d(t_{2g})$ levels of the manganese fall below the doubly occupied oxygen 2p levels corresponding to an energy separation of ~10 eV between the d orbitals and the top of the valence band states. At first glance this situation seems to reflect a highly excited electronic state and one might guess that tetravalent manganese is extremely unstable against charge transfer. However, two important facts should be borne in mind. First, the the eigenvalues of (small) clusters are not simply related to crystalline eigenvalues[4], which is plain from the discussion leading to (2.51). Thus the cluster eigenvalues do not provide information on the position of defect levels within the band gap. The second remark concerns the physical significance of orbital energies of small quantum systems such as atoms or molecular clusters. Within HF theory it is Koopmans' theorem which identifies the orbital energies with negative ionization energies. Orbital energy differences are then approximate excitation energies. However, there is a precondition for this theorem to work in practice, i.e. excitation-induced orbital relaxation effects must be negligibly small. Whereas this condition is fulfilled for delocalized eigenstates of large systems like crystals, it is not guaranteed for small clusters of atoms. As an example we consider the present $(MnO_6)^{-8}$ cluster. First, the $^4A_{2g}$ electronic state of this cluster has been calculated self-consistently. Then, without allowing any further orbital relaxations, the total energies of the two ionized states with one electron removed from either the top of the oxygen 2p levels or from the $d(t_{2g})$ manganese orbitals have been determined. The total energy differences with respect to the $^4A_{2g}$ state provide the corresponding ionization energies. On the basis of this procedure the precondition of Koopmans' theorem has been forced to be operative. The difference between the ionization energies obtained in this way (~10 eV) equals the difference of orbital energies quoted above. Next, the calculation of the ionization energies was repeated, but taking into account all orbital relaxations. As a result of orbital relaxations (which mainly affect the ionization from the localized d orbitals, corresponding to an energy gain of ~7.1 eV) the difference in the ionization energies for the two processes reduces to about 3.9 eV. Thus, the ΔSCF method shifts the $d(t_{2g})$ related states of manganese much closer to the oxygen 2p states. Related discussions of the breakdown of Koopmans' theorem in transition metal complexes are reported in [87, 88, 89].

It is emphasized that ΔSCF energies may be considered as simplified cases of total crystal energy differences calculated on the basis of (2.57), if the individual total energies differ only with respect to the cluster state function Ψ.

[4] By extending the cluster size to infinity one would certainly observe that the cluster orbitals (and energies) converge to their bulk limits. This relation provides information on how fast the cluster orbitals approach the delocalized crystalline eigenstates. Typically one must consider large clusters in order to obtain the properties of delocalized bulk orbitals. This observation, however, does not invalidate (small) embedded cluster calculations, because their justification derives from the use of localized crystalline orbitals based on (2.45).

In this fashion, differences between SCF energies are the physically relevant quantities which are consistent with the embedded cluster approach, not orbital energy differences. The situation differs slightly in the case of embedded cluster calculations based on the Xα exchange approximation [18] or, more generally, on the density functional theory (DFT). Again, *a priori*, the cluster orbital energies are not related to crystalline one-electron energies. But different to HF, there is from the beginning no formal basis for Koopmans' theorem relating the one-electron energies to (unrelaxed) ionization energies. Instead, the theorem of Slater [18] (generalized for DFT by Janak [90]) holds:

$$\varepsilon_i = \frac{\partial E[\varrho]}{\partial n_i} \,. \tag{2.59}$$

For large electron numbers this theorem provides a relation between orbital energies ε_i and corresponding occupation numbers n_i according to the Fermi–Dirac statistics. It is, thus, plain that the different meanings of HF and Xα one-electron energies may result in a substantially different level ordering. As an example we compare the HF orbital energies of $(MnO_6)^{-8}$ with the corresponding ones obtained from earlier Xα cluster calculations [91]. Whereas within HF the $d(t_{2g})$ manganese orbital energies are well separated from the oxygen 2p levels, the same d orbital energies are close to the top of the oxygen 2p states in the case of the Xα cluster calculations. Following Slater [18], the differences between the orbital energies can be traced back to the differences in the respective exchange potentials:

$$\varepsilon_i^{\mathrm{X}\alpha} - \varepsilon_i^{\mathrm{HF}} = V_{\mathrm{X}\alpha}(\boldsymbol{r}) - V_{\mathrm{XHF},i}(\boldsymbol{r}) \,. \tag{2.60}$$

The inspection of Slater–Condon exchange parameters of transition metal atoms suggests that in particular the 3p–3d exchange interaction is responsible for the expected differences in the exchange potentials.

Both the HF and the Xα one-electron energies are qualitatively related to an ionization, i.e. the first ones on the basis of Koopmans' theorem and the latter ones as a consequence of the Fermi–Dirac statistics (the highest occupied orbitals should most easily be ionizable). It may be assumed that the Xα and HF orbital energies represent lower and upper bounds on the true ionization energy based on ΔSCF, which, however, can be substantially different from the orbital energies, if the difference on the left-hand side of (2.60) is large. Then we have to expect considerable orbital relaxation effects. Only in situations of delocalized orbitals can we expect that the difference between the Xα and HF exchange potentials will be sufficiently small. The corresponding agreement of orbital energies then leads to a physical interpretation of the one-electron energies, since both Koopmans' theorem and the Fermi–Dirac condition are approximately fulfilled. This situation corresponds to crystalline one-electron eigenstates close to the Fermi energy and agrees with the general perceptions of DFT.

2.2 Potential Simulations

2.2.1 Potential Forms and Applications

Generally, potential simulations of crystals are based on the validity of (2.5). In this section various functional forms of crystal potentials and some of their applications are reviewed. The derivation of crystal potentials is considered in Sect. 2.2.2. Major emphasis will be placed on pair potential descriptions, since these have proved to simulate reliably the formation of defects in many oxide systems of technological interest. The subsequent chapters are mainly devoted to corresponding questions. More general potential forms are useful only if one needs to model accurately the minute structure and energy changes due to FE-PT in perovskite oxides.

The effective crystal potential V_{eff} may be obtained by evaluating $V_{nn}(\mathcal{R}_i)$ + $E_0(\mathcal{R}_i)$ for a set of lattice configurations $\{\mathcal{R}_i\}_{i=1}^{n}$. Potential energies are then fitted with a general *model potential* expression defined with respect to the nuclear coordinates or to appropriately chosen symmetry-adapted coordinates. In many situations symmetry-related restrictions are necessary in order to reduce the number of independent variables to a manageable level. Avoiding any approximating assumptions with respect to the potential form, the potential energy surface (PES) of the whole crystal can only be scanned at specified high-symmetry points within the first Brillouin zone by employing suitable phonon-adapted ion displacements. Such an approach, based on model potentials derived from perfect crystal *ab initio* calculations, is often employed to investigate general phonon-related structural transitions, of which ferroelectric instabilities in perovskite oxides (see also Sect. 3.1) provide examples. Within the applied *ab initio* level model potentials provide an accurate image of the Born–Oppenheimer PES of the ground state.

A brief illustration of these ideas can be given by sketching the construction of an effective lattice Hamiltonian in terms of (localized) lattice-type Wannier displacements. In this group theoretical approach [92] the site symmetry concept is employed (see Sect. 2.1). One proceeds by decomposing the $3n_{at}N$-dimensional "ionic-displacement" space (with n_{at} ions contained in each of the N unit cells) into a direct sum of invariant subspaces (so-called "band subspaces"). These band subspaces are spanned by one or more entire branches of normal modes and are closed under the action of the space group $\mathcal{G}_0$, belonging to a high-symmetry crystal structure to which the particular structural transition refers. This decomposition corresponds to the band picture of the perfect crystal electronic structure. For each band subspace the constructed set of localized Wannier displacement basis vectors (the lattice analogue of the electronic Wannier orbitals, Sect. 2.1) transform as partners of a particular irreducible representation of the site symmetry group corresponding to one specified Wyckoff position index. The complete lattice Hamiltonian is written in terms of the Wannier displacements (i.e. the coordinates with respect to the localized Wannier basis vectors) by performing a Taylor expansion with respect to the reference structure. The coefficients of this expansion may be obtained from *ab initio* calculations. At quadratic order there are no cross terms between the band subspace Λ_0 in which the effective symmetry-adapted Hamiltonian being sought acts and all other band subspaces $\Lambda \neq \Lambda_0$. One finds:

$$V_{\text{cryst}} \approx V_{\text{eff}}(\{\xi_{\Lambda_0 l \boldsymbol{R}_n}\}) + \sum_{\Lambda \neq \Lambda_0} V_{\Lambda}^{(2)}(\{\xi_{\Lambda l \boldsymbol{R}_n}\}) \,. \tag{2.61}$$

The $\xi_{\Lambda l \boldsymbol{R}_n}$ denote the Wannier displacements. The labels refer to the respective band subspace Λ, the normal modes contained in Λ and the appropriate Wyckoff positions in the lattice. The effective Hamiltonian may be employed in statistical mechanical analyses. Based on the decoupling of band subspaces, the relevant partition function depends only on the set $\{\xi_{\Lambda_0 l \boldsymbol{R}_n}\}$. Obviously the described procedure reduces the number of variables to a manageable amount.

In an alternative approach the effective crystal potential is expanded in terms of multi-body interactions employing the coordinates of the crystalline nuclei:

$$V_{\text{eff}}(\mathcal{R}) = V_0 + \frac{1}{2!}\sum_{i,j}{}' V_{ij}(\boldsymbol{R}_i, \boldsymbol{R}_j) + \frac{1}{3!}\sum_{i,j,k}{}' V_{ijk}(\boldsymbol{R}_i, \boldsymbol{R}_j, \boldsymbol{R}_k) + \ldots, \tag{2.62}$$

where primes denote the exclusion of equal indices in the summations. The configuration-independent term V_0 (sum of atomic or ionic self-energies) is often neglected. In order to characterize the global PES of a solid the higher multi-body terms must be considered, too. Termination of the expansion in (2.62) principally restricts the applicability of the potential to parts of the crystalline configuration space. However, in many situations the second and third summands provide the dominating contributions. The second term in (2.62) describes pair potentials and the third three-body contributions. In the valence force model (see [93]) of covalent solids these two summands are used to simulate the stretching and bending of covalent bonds close to their ideal structure. The model may be applied to calculations of perfect-lattice and defect problems. Scheffler et al. [94] applied an *ab initio* derived valence force potential to investigate the formation of the Ga vacancy in GaAs.

Another important simplification of (2.62) consists of employing the pair potential central field approximation (PP-CFA):

$$V_{\text{eff}}^{\text{PP-CFA}}(\mathcal{R}) = V_0 + \sum_{i<j} V_{ij}(|\boldsymbol{R}_i - \boldsymbol{R}_j|) \,. \tag{2.63}$$

Potentials of this type are useful in simulations of van der Waals bonded crystals, ionic systems and even of simple metallic crystals, where the nearly free electron model works accurately [95].

Potential descriptions going beyond the pair and three-body potential approaches refer to pair and cluster functionals [96]. Such generalized potentials efficiently add environment dependent terms to the pair and cluster potentials, thereby extending the covered part of the configuration space. Corresponding functionals are necessary to account for transition metals. In semiconductors these contributions improve on simulating broken bonds. A pair functional, for example, reads as:

$$V_{\mathrm{eff}}^{\mathrm{PF}}(\mathcal{R}) = \frac{1}{2!}\sum_{i,j}{}' V_{ij}(\boldsymbol{R}_i, \boldsymbol{R}_j) + \sum_i U\left(\sum_j g(\boldsymbol{R}_i, \boldsymbol{R}_j)\right) . \tag{2.64}$$

Whereas the first term in (2.64) denotes a pair potential, the second summand defines a functional assigning a real number to the environmental pair function g. The function U is generally non-linear; otherwise (2.64) fits into the pair potential scheme. The formal justification of pair functionals can be based on a tight binding model analysis or on an embedded-atom approach. Whereas in the first case g refers to the local band width (i.e. the second moment of the site-projected density of states) and U obeys a square root dependence, g relates to the environment electron density of an atom and U is a numerical function in the latter case. Pair functionals have been applied to investigations of intrinsic defect formations including broken bonds and energy differences between competitive crystal structures.

Many potential simulation studies of insulating solids employ the PP-CFA extended by appropriate electronic polarization contributions. The shell model represents the most successful scheme in this respect. Originally shell model-based simulations were invented to investigate highly ionic materials (e.g. alkali halides). Later, calculations of this type turned out to be extremely successful also for many complex oxides – even in systems with pronounced covalency, such as silicates. The shell model approach also provided detailed information on the defect chemistry of electro- and magnetooptic oxides (for a review see the subsequent chapters) as well as of high-T_c copper-based superconductors (La_2CuO_4 [97, 98, 99], $YBa_2Cu_3O_{7-\delta}$ [100, 101, 102, 103, 104] and $HgBa_2Ca_2Cu_3O_{8+\delta}$ [105]). Because of this success it is useful to summarize briefly the fundamentals of the shell model and its applications.

The effective potential is given in the pair potential central field approximation (PP-CFA). Higher-order corrections (i.e. three-body bond-bending terms) may be added where they appear to be necessary. Electronic polarizations of the crystal ions are described on the basis of (an)isotropic spring constants in harmonic or even anharmonic approximation. The spring constant couples a shell to each ionic core, which reads in the isotropic cases as:

$$V_{\mathrm{core-shell}}(\boldsymbol{r}_{\mathrm{core}}, \boldsymbol{r}_{\mathrm{shell}}) = \frac{1}{2}k|\boldsymbol{r}_{\mathrm{core}} - \boldsymbol{r}_{\mathrm{shell}}|^2 . \tag{2.65}$$

The electronic polarizability arising from this core–shell interaction is given by the expression:

$$\alpha = \frac{(q_{\mathrm{shell}})^2}{k} , \tag{2.66}$$

where q_{shell} denotes the shell charge of the respective ion. Total ionic charges $q = q_{\text{core}} + q_{\text{shell}}$ may be chosen as formal integral charges corresponding to an ideal ionic crystal model or as partial charges. The actual V_{SR}^{ij} depend on this choice. It is noted that the transferability properties of short-range potentials goes wrong for partial charge models. Phonon calculations often invoke partial charges in order to obtain a better representation of dispersion curves. Lattice energies, on the other hand, are more satisfactorily calculated with formal charges. This is also true for defect formation energies, which are differences between perfect and defective lattice energies. Moreover, it is straightforward to formulate defect chemical reactions using formal ion charges, but not so with partial charges.

Considering (semi-)ionic systems, the sum of pair potentials in (2.63) is split into the long-range (LR) Coulomb part and short-range (SR) potentials describing electronic repulsion and covalency effects. If one is mainly interested in modelling non-interacting phonons, it suffices to apply the harmonic approximation to (2.63). However, in all cases in which large ion displacements are to be expected (e.g. around defects with pronounced charge or size misfit) the harmonic approximation may lead to inaccuracies. Then it is preferable to employ the explicit PP-CFA interionic potentials:

$$V_{\text{LR}}^{ij}(r_{ij}) = \frac{q_i q_j}{r_{ij}} . \tag{2.67}$$

q_i and q_j are core or shell charges of different ions and r_{ij} is the interionic separation. Short-range pair potentials are conveniently chosen as Buckingham potentials, i.e.

$$V_{\text{SR}}^{ij}(r_{ij}) = A_{ij} \cdot \exp\left(-\frac{r_{ij}}{\varrho_{ij}}\right) - \frac{C_{ij}}{r_{ij}^6} . \tag{2.68}$$

Other choices (e.g. Morse potentials) may be used as well. The first term appearing on the right-hand side of (2.68) represents repulsive interactions arising from the impenetrability of closed electron shells; the second (attractive) term models the van der Waals interactions but also covalency. Arising from electron–electron interactions, V_{SR}^{ij} should be defined as acting between ion shells. This modifies the in-crystal polarizabilities of ions and effectively makes them dependent on the host – an important effect, particularly for the diffuse oxygen anions. Effectively, the corresponding changes can be introduced as

$$\alpha_{\text{in-crystal}} = \frac{(q_{\text{shell}})^2}{k + k_{\text{SR}}} , \tag{2.69}$$

where k_{SR} results from the short-range interactions between neighbouring ions. Figure 2.3 schematically displays the various types of interaction within a shell model description. The shell model is superior to unpolarizable (or rigid) ion models, because it may correctly account for the high-frequency dielectric properties of crystals. It also improves on polarizable point ion (PPI)

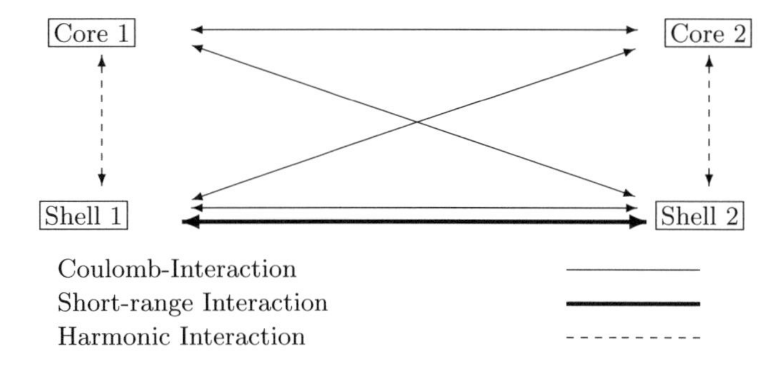

Fig. 2.3. Pair potential interactions within a shell model description of crystals

models due to the coupling of ion polarizations and short-range interactions. Within shell model simulations polarization catastrophes can be successfully suppressed. For further details the reader is referred to [106]. More elaborate versions of the simple shell model discussed so far consider the deformation of shells (breathing shell model). Such models are able to account for observed Cauchy violations; defect energies, on the other hand, do not change significantly upon these improvements [107, 108]. The general problems with elaborate shell models refer to a rapid rise in the number of potential parameters to be determined.

Having developed a suitable parametrization of shell model-based potentials (see Sect. 2.2.2), the lattice structure and phonons of perfect crystals may be investigated. It is also possible to calculate all macroscopic crystal properties which are in the realm of suitable potential models. Corresponding properties include the elastic and dielectric constants, for instance. Green's function methods, supercell simulations or Mott–Littleton-type calculations can be used to study the structural and energetical properties of defects. The Mott–Littleton approach considers isolated point defects or small aggregates and employs a two-region strategy to obtain the required internal defect formation energies u_{v} (see Fig. 2.4). Outer crystal regions are treated as a polarizable continuum.

The basic ideas of Mott–Littleton formulations may be summarized as follows (for details see [106]). The internal defect formation energies can be written as:

$$\begin{aligned} u_{\mathrm{v}} &= \min_{\Psi,x,y}\{E(\Psi,x,y)\} \\ &= \min_{\Psi,x,y}\{E_{\mathrm{I}}(\Psi,x) + E_{\mathrm{I-II}}(\Psi,x,y) + E_{\mathrm{II}}(y)\}\,. \end{aligned} \tag{2.70}$$

The defect energy consists of contributions due to region I and II and the interaction term $E_{\mathrm{I-II}}$ (see also Fig. 2.4 for additional information); x and y denote the totality of region I and II ion displacements, respectively. The minimization with respect to x and y includes defect-induced lattice relaxations.

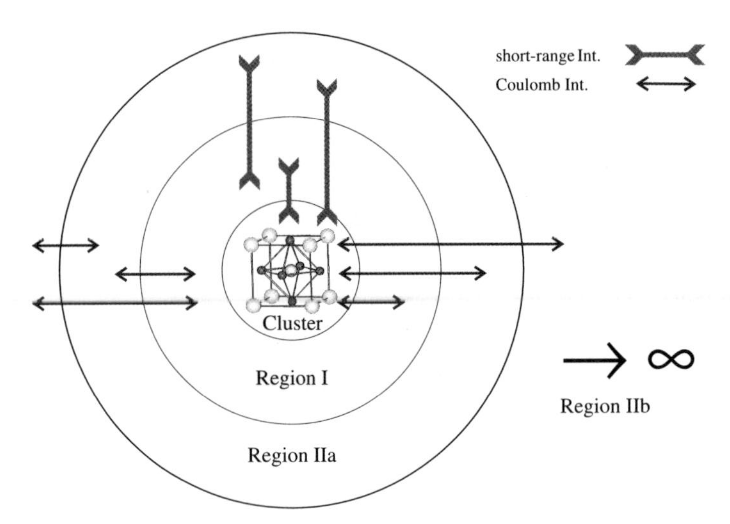

Fig. 2.4. The two-region strategy. The (embedding) crystal lattice is divided into the regions I, IIa and IIb. The inner region I which immediately neighbours the respective defect (simulated by the central defect cluster in this figure) usually contains between 100 and 300 ions. Region I is explicitly equilibrated according to the underlying pair potentials. The outer region II (consisting of parts a and b), on the other hand, is treated as a polarizable continuum using the harmonic approximation for the region II self-energy. Whereas the interactions between the cluster and region I, on the one hand, and region IIa, on the other hand, are included explicitly, all region IIb species feel only the effective defect charge of the central defect cluster. Arrows denote the relevant interactions between the various regions

The defect energies in (2.70) are internal energies at constant volume, because in practice any elastic deformations of the embedding region II are neglected. Thus the volume of the crystal remains unchanged upon defect formation. The defect wavefunction Ψ has been introduced to account for embedded cluster calculations; for corresponding details the reader is referred to Sect. 2.1.2. In the present context we restrict our discussion to considering the classical degrees of freedom x and y. The remaining minimization in (2.70) becomes reduced to a finite problem by employing the constrained search condition,

$$u_{\mathrm{V}} = \min_{x} E(x, y_0(x)) := \min_{x}(\min_{y} E(x, y)), \tag{2.71}$$

and the harmonic approximation applied to region II:

$$E_{\mathrm{II}}(y) = \frac{1}{2} y^{\mathrm{T}} A y . \tag{2.72}$$

Effectively, (2.71) assumes region II to be adiabatically equilibrated for each arbitrarily fixed region I configuration. The Mott–Littleton approach employs dielectric continuum theory in order to approximate the unknown functional dependence $y_0(x)$. For example, considering an isotropic ionic crystal with two unpolarizable ions (charge $\pm q$) per unit cell (volume V_{uc}), the defect-induced relaxations of distant region II ions are given by:

$$\boldsymbol{y}_{0,+q} = -\boldsymbol{y}_{0,-q} \approx \frac{V_{\mathrm{uc}}}{8\pi|q|}\left(1-\frac{1}{\varepsilon}\right) Q \frac{\boldsymbol{r}}{r^3} . \tag{2.73}$$

ε denotes the dielectric constant of the material and Q the effective defect charge. Because of the factor $(1-\frac{1}{\varepsilon})$, the precise value of the dielectric constant is of little importance once it becomes large. Also, the effect of dielectric anisotropy will be negligible in such cases. Equations (2.71-2.73) can be used to calculate the asymptotic contributions to $E_{\mathrm{I-II}} + E_{\mathrm{II}}$:

$$E_{\mathrm{I-II}} + E_{\mathrm{II}} \approx -\frac{1}{2} Q \sum_{j \in \mathrm{II}} q_j \frac{\boldsymbol{y}_j . \boldsymbol{r}_j}{r_j^3} . \tag{2.74}$$

The infinite summations of Coulomb contributions in (2.74) may be carried through exactly using the Ewald technique. Finally, the required minimization procedure reduces to the finite problem:

$$\left.\frac{\partial E(x,y)}{\partial x}\right|_{y=y_0} = \frac{\partial}{\partial x}\{E_{\mathrm{I}}(x) + E_{\mathrm{I-II}}(x, y_0)\} = 0 . \tag{2.75}$$

This explicit minimization of the embedded region I can be conveniently performed by employing the variable metric technique [81].

The formal minimization scheme extended to complex and anisotropic crystals has been implemented in a number of computer codes, e.g. HADES [106], CASCADE [109] and GULP [110].

Though it is possible to calculate the entropy contributions of isolated defects (i.e. vibrational contributions) to the free energy of defect formation $f_{\mathrm{v}} = u_{\mathrm{v}} - Ts_{\mathrm{v}}$ [10], it turns out that in many instances the entropy term $(-Ts_{\mathrm{v}})$ is sufficiently small compared with the internal formation energy u_{v}. Corresponding results found in ionic crystals [10] are certainly useful in oxides, too. Significant entropy terms may enter the calculations only, if gas-phase states are involved. Nevertheless, for predictions related to the defect chemistry of materials it will often suffice to calculate internal defect-formation energies u_{v} and to combine them according to appropriate defect reactions.

For a general discussion of well-known simulation techniques, see the extensive review by Catlow and Mackrodt [106] and the recent review by Harding [10]. These references also provide detailed information on surface simulations, molecular dynamical (MD) and Monte Carlo statistical calculations. It is finally emphasized that particularly MD-simulations become feasible within a pair potential approach. Calculations of this type have been performed to investigate the dynamics of FE-PT in perovskite oxides (see Sect. 3.1). In particular, for atomistic simulations of structure, energetics and impurity segregation at oxide surfaces the reader is referred to [111]. Note that these investigations also invoke Mott–Littleton type methods.

2.2.2 Derivation of Effective Crystal Potentials

In practice the generation of crystal potentials employs suitable fitting procedures. The first (theoretical) approach is based on fitting the calculated potential energy surface (PES) of either the total crystal or an appropriately chosen embedded molecular fragment. In the second (empirical) method the chosen parameters of the potential are adjusted to reproduce the observed crystal structure and macroscopic properties (e.g. elastic and dielectric constants or vibrational properties). The following discussion is restricted to considering the derivation of shell model pair potential parameters, but similar procedures can be developed as well for more general potentials.

Least-square fitting procedures invoke the minimization of a chosen cost function which in theoretical approaches reads as:

$$S_{\mathrm{t}} = \sum_i w_i |E_{\mathrm{PES}}(\mathcal{R}_i) - V(\mathcal{R}_i; \{\alpha\})|^2 \tag{2.76}$$

with (see also Sect. 2.1)

$$E_{\mathrm{PES}}(\mathcal{R}_i) = V_{nn}(\mathcal{R}_i) + E_0(\mathcal{R}_i) \,. \tag{2.77}$$

In (2.76) the sum runs over the specified ion configurations $\mathcal{R}_i$ of the considered system, w_i denotes weight factors possibly defined, V is the required potential and $\{\alpha\}$ is the respective set of potential parameters which in the framework of the shell model includes A, ϱ, C for each (Buckingham) pair potential and Y, k for each shell model ion. In highly ionic materials (the alkali halides or magnesium oxide, for example) the quantum clusters can be assumed to consist of pairs of ions of which the short-range interaction is required. In order to determine E_{PES} of such dimers one can employ either *ab initio* cluster calculations [112, 113, 114] or the electron gas procedure according to Gordon and Kim [115], of which the latter approach is based on the LDA ground state energy functional of closed-shell atoms. Generalizations to open-shell atoms may be inferred from [116].

In the electron gas method the short-range potential between ions A and B is derived by investigating

$$E_{\mathrm{PES}}(\mathrm{AB}) = E[\varrho_{\mathrm{A}} + \varrho_{\mathrm{B}}] - E[\varrho_{\mathrm{A}}] - E[\varrho_{\mathrm{B}}] \tag{2.78}$$

as a function of the interionic separation between A and B. The assumed rigid superposition of ionic densities leads to the neglect of any covalency effects. The method is particularly useful in describing highly ionic crystals. In simulation studies of semi-ionic materials, electron gas derived potentials are likely to overestimate the repulsive interaction contributions. It is an advantage of the electron gas method that it does not mix up the short-range interionic potentials and the electronic polarizations of the ions. Any shell parameters required must be inferred from additional empirical fitting to dielectric properties of the considered material [117].

Ab initio cluster calculations, on the other hand, are applicable to all types of material in order to derive interatomic (interionic) potentials. The investigated clusters may consist of ion pairs where this appears to be appropriate, but choosing larger clusters should be preferred in most situations. The necessity for extended clusters relates to complex covalency effects (see the derivation of an *ab initio* niobium–oxygen potential in $KNbO_3$, below). Another sophisticated approach towards the derivation of *ab initio* pair potentials has been suggested recently by Recio et al. [118]. In this atom-in-crystal method, based on the group function description of McWeeny and Huzinaga (Sect. 2.1.2) the lattice energy of the crystal is written as a sum of atomic (ionic) self-energies and atomic (ionic) interaction contributions in the crystal. The method, which is applicable if the relevant group functions can be chosen as atomic(ionic)-in-crystal wavefunctions, has been tested for different alkali halides but not for complex oxides.

The particular derivation of impurity-related potentials has been recently addressed by Pandey et al. [119]. The corresponding approach employs an embedded cluster description including defect-induced relaxations of the outer lattice spheres (see Sect. 2.1.2). In particular, for charged defects lattice relaxations are able to suppress unwanted charge oscillation effects.

Ab initio cluster calculations require an appropriate embedding of the chosen quantum cluster to mimic the crystalline environment (Sect. 2.1.2). Most importantly, the Madelung potential must be included. The more diffuse the crystalline anions are, the more important becomes the inclusion of the Madelung potential [113, 117, 120]. This is particularly obvious for oxides, because O^{2-} anions are unstable as isolated species. The stabilizing influence of the lattice potential is indispensable in this case. Recent computer modelling of the crystalline boric oxide B_2O_3 [121] has indicated significant failures of potentials derived from *in vacuo* quantum cluster calculations. It turned out to be insufficient to add the Madelung potential afterwards.

By construction, *ab initio* cluster calculations mix interionic potentials with polarization contributions of the ionic charge densities. The resulting potentials may not be used in shell model simulations, as this would introduce artificial double-counting effects otherwise. For pairs of ions, Harding and Harker [113] suggested a procedure based on a quadrupole analysis to separate the potential and polarization contributions. In the case of complex clusters one may conveniently employ hydrostatic pressure simulations in order to derive polarization-free *ab initio* pair potentials (see the derivation of an *ab initio* niobium-oxygen potential in $KNbO_3$, below). Additional information on polarizations can be obtained empirically by fitting to dielectric properties or by further cluster calculations involving symmetry-breaking distortions.

Finally, it is possible to fit the calculated PES of the perfect crystal ("infinitely large cluster") with model potentials. This approach has been recently applied to B_2O_3 [121]. The linear programming methodology employed (see also [81]) has been designed to fit the PES of the system with the additional constraint of requiring the reproduction of observed crystal structures. Both the crystal energy to be optimized and the structural constraints have been taken linearized with respect to the adjustable potential parameters. Because no care was taken to distinguish between the interionic potential and ionic polarization contributions, the resulting set of pair potentials delivers a rigid ion model. As in the case of cluster calculations, hydrostatic pressure simulations supplied with additional empirical or theoretical investigations would remedy this restriction.

Empirical derivations of potentials are based on fitting procedures to the observed crystal structure and macroscopic properties. Considering oxides, this easy-to-handle methodology is the most frequently used strategy to develop suitable crystal potentials, although theoretical techniques are conceptually more desirable. Basically the empirical cost function can be written as:

$$S_{\mathrm{e}} = \sum_{i} w_{1,i}\varepsilon_i^2 + \sum_{j,k} w_{2,jk}\varepsilon_{jk}^2 + \sum_{l} w_{3,l}(P_{\mathrm{calc}}(l) - P_{\mathrm{exp}}(l))^2 . \tag{2.79}$$

The lattice and basis strains, ε_i and ε_{jk}, usually refer to the observed crystal structure. The entities $P_{\mathrm{calc/exp}}(l)$ denote calculated and measured macroscopic crystal properties, respectively. These depend on the matrix of second derivatives of the lattice energy and, thus, provide some information on the effect of changing interionic separations. Empirical potentials are by construction strictly valid only close to the observed lattice spacings. Consequently, these potentials may fail to work in situations with significantly deviating interionic separations (e.g. "extreme" interstitial defect problems). For complex materials it may happen that the number of fit parameters exceeds the number of known experimental data. In this situation it is often helpful to fit related binary systems and transfer the corresponding potentials to simulations of the more complex crystals. Quite generally, testing the transfer of potentials may help to control the physical interpretability of derived potential sets. Further, it is common practice to employ the transfer method in order to simulate impurities. Detailed information on empirical potentials for oxides can be inferred from [12, 122].

There are several strategies to improve empirical potential derivations. First, (2.79) can be easily generalized to perform a simultaneous fit of multiple crystal structures and their properties (e.g. different structural modifications of one crystal) [110]. Second, one can employ relaxed fitting procedures [110] in which the calculated lattice and basis strains as well as macroscopic properties are taken with respect to the particular crystal structure being fully equilibrated according to the actual potential parameters. The relaxed fitting technique is especially useful for materials that are close to structural instability. Oxide perovskites, which are at least incipient ferroelectrics, represent important examples. The advantage of relaxed fitting routines derives from the observation that minute changes of ion positions, which occur when residual bulk and basis strains are released after conventional fitting, may result in pronounced changes of material properties like the dielectric constants.

The method has been recently applied to simulations of the incipient ferroelectric $KTaO_3$ (see Chap. 4). Further, using the relaxed fitting approach, progress has been made towards stabilizing the ferroelectric tetragonal phase of $PbTiO_3$ over the cubic phase which is usually generated within isotropic shell model simulations [123]. Third, the inclusion of oxygen core shell polarization vectors as adjustable parameters in the fitting procedure may help to stabilize required ferroelectric phases. This method has been successfully applied to model the ferroelectric R3c phase of $LiNbO_3$ [124].

Nb^{5+}... O^{2-} Short-Range Pair Potentials Derived from Embedded Cluster Calculations in $KNbO_3$. This subsection is devoted to delineating further aspects of *ab initio* derivations of pair potentials. In particular, we consider the derivation of *ab initio* Nb^{5+}... O^{2-} short-range pair potentials which are obtained from *ab initio* SCF-LCAO-MO embedded cluster calculations. The underlying crystal structure refers to cubic $KNbO_3$. Two types of cluster are considered in order to derive the short-range niobium–oxygen interaction, i.e. a simple $Nb^{5+}O^{2-}$ pair and a more complex $(NbO_6)^{7-}$ cluster. Both fragments are embedded in an appropriate set of point charges. Figure 2.5 shows the $(NbO_6)^{7-}$ cluster with its nearest potassium and niobium point charge neighbours.

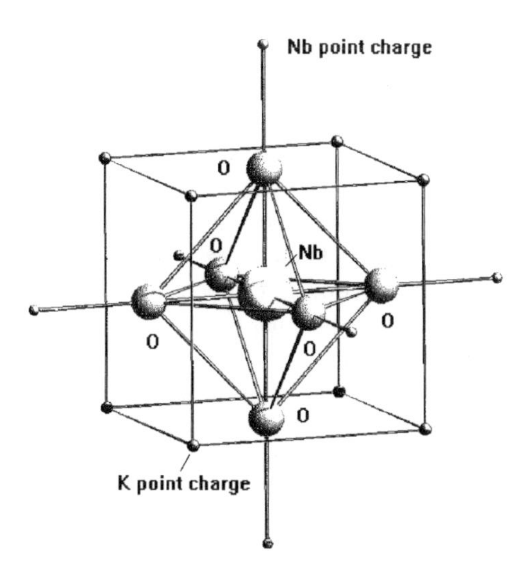

Fig. 2.5. $(NbO_6)^{7-}$ quantum cluster with its nearest potassium and niobium (point charge) neighbours. Spheres do not correspond to known ion radii

The basis functions are chosen to be the split valence SV-21 set of Huzinaga [125] for Nb and the set of Dunning and Hay [126] for the oxygen ions. The oxygen basis set is modified by completely breaking down the p-contractions and by augmenting with diffuse p- and polarizing d-type functions. As a consequence of this improved basis set quality we can expect that possible basis set superposition errors [127] are substantially smaller than the corresponding errors estimated for oxygen ions in MgO using a TZV basis set (1.5eV [128]).

It is one of our aims to compare the *ab initio* potentials with a given empirical Nb^{5+}... O^{2-} potential. As the latter has been fixed to be a pure Born–Mayer exponential function [12], it is reasonable to employ the Hartree–Fock approximation without configuration interaction (CI) in all subsequent cluster calculations and to fit the results with a Born–Mayer potential as well.

Short-range potential energies may be calculated for various cluster geometries according to the following formula:

$$\begin{aligned} V_{\mathrm{SR}}^{\mathrm{Nb}^{5+} \ldots\, \mathrm{O}^{2-}} &= \frac{1}{n_{\mathrm{NbO}}} \Big[E_{\mathrm{c}}^{\mathrm{HF}} - \sum E_{\mathrm{self}}^{\mathrm{HF}} - E_{\mathrm{cb}} \\ &\quad - \sum V_{\mathrm{SR}}^{\mathrm{O}^{2-} \ldots\, \mathrm{O}^{2-}} - E_{\mathrm{pol}} \Big] . \end{aligned} \tag{2.80}$$

n_{NbO} denotes the number of Nb–O bonds in the cluster, $\mathrm{E}_{\mathrm{c}}^{\mathrm{HF}}$ the quantum mechanical cluster energy, $\sum \mathrm{E}_{\mathrm{self}}^{\mathrm{HF}}$ the sum of ionic self-energies, E_{cb} the rigid ion Coulomb contribution according to the formal charge model, $\sum V_{\mathrm{SR}}^{\mathrm{O}^{2-} \ldots\, \mathrm{O}^{2-}}$ the occurring oxygen–oxygen short-range interactions and E_{pol}, finally, an ionic polarization term. Ionic self-energies $E_{\mathrm{self}}^{\mathrm{HF}}$ are extracted by treating one ion at a time quantum mechanically and the other ions as point charges consistent with the respective cluster geometry. In order to be consistent with the empirical $KNbO_3$ potential model, $\sum V_{\mathrm{SR}}^{\mathrm{O}^{2-} \ldots\, \mathrm{O}^{2-}}$ in (2.80) has been retained from that model. It is noted that Catlow et al. [129] derived this anionic interaction as well from *ab initio* calculations. Since then it has been successfully applied in many simulation studies of oxides. Thus there is no apparent need to replace this potential. To avoid double-counting effects when using *ab initio* short-range potentials in shell model simulations, polarization terms must either be subtracted from these potentials, as was done in (2.80), or a procedure should be devised where electronic polarization of ions does not occur. E_{pol} in (2.80) consists of three different parts, all of which result from core–shell polarizations, i.e. they arise from transforming from a rigid ion to a shell model description:

$$E_{\mathrm{pol}} = E_{\mathrm{pol}}^{\mathrm{cb}} + E_{\mathrm{pol}}^{\mathrm{Harm}} + E_{\mathrm{pol}}^{\mathrm{SR}} . \tag{2.81}$$

$E_{\mathrm{pol}}^{\mathrm{cb}}$ denotes corrections to E_{cb} defined above, $E_{\mathrm{pol}}^{\mathrm{Harm}}$ the harmonic polarization energy and $E_{\mathrm{pol}}^{\mathrm{SR}}$ all corrections to short-range interactions. This last term results from the differences between core and shell positions; it is recalled that short-range potentials were defined to act between the ionic shells.

In case of the single $Nb^{5+}O^{2-}$ pair we consider only oxygen ion displacements along the [100] cubic crystal direction parallel to the molecular pair. In Fig. 2.6 the corresponding short-range Nb^{5+}... O^{2-} potentials are shown

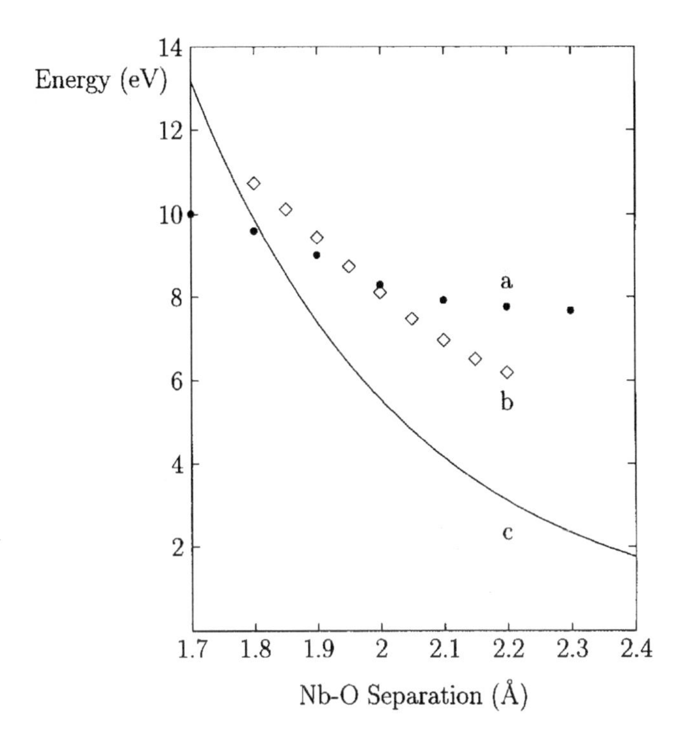

Fig. 2.6. *Ab initio* potential derived from a simple pair of ions. (a) *Ab initio* potential (•) from a $Nb^{5+}O^{2-}$quantum pair without polarization correction. (b) *Ab initio* potential (◇) from a $Nb^{5+}O^{2-}$quantum pair including polarization correction. (c) Empirical Nb^{5+}... O^{2-} short-range potential

with and without polarization correction. The polarization correction has been estimated by considering the additional influence of polarizing oxygen basis functions. Importantly, these functions do not improve the oxygen self-energy. It is further noted that the expected polarization contributions are mainly due to the highly diffuse oxygen anion; the niobium cation, on the other hand, may be considered as being rigid to first order.

Whereas simple pair cluster models account sufficiently for purely ionic crystals (e.g. [106, 112]), they provide completely unsatisfactory crystal descriptions for more covalent materials like $KNbO_3$. This result is independent of the inclusion of the polarization term. Figure 2.6 clearly demonstrates the pronounced differences between the present *ab initio* potential dependence and the corresponding empirical potential, which may be taken as a measure of quality due to its successful simulation of the Nb...O interaction in $KNbO_3$.

In the case of the $(NbO_6)^{7-}$ quantum cluster we consider two different sets of calculations. The first set relates to breathing mode distortions of the octahedral cluster within an otherwise fixed point-ion lattice. Except for the perfect lattice spacing we are faced with the problem of electronic polarization of the oxygen ions. Principally this problem can be understood on the basis of symmetry arguments[5]. However, there is no practical prescription to obtain appropriate quantitative information on polarizations from an analysis of the charge density. Since all cluster multipole moments vanish up to hexadecapole moments, the method proposed by Harding and Harker [113] is not useful in this case. Instead, we devise in the second approach a hydrostatic-pressure simulation, which means that the total crystal is shrunk or expanded corresponding to a change of the cubic lattice constant. Thus, at each Nb–O separation the system maintains perfect cubic lattice symmetry. As a result there are essentially no electronic polarization effects and, thus, the short-range potential obtained in this way is not contaminated by polarizations. The hydrostatic pressure approach has physical significance, since it tests by construction the pressure dependence of the NbO_6 crystal fragment. It is general and may be applied to all *ab initio* schemes in order to derive crystal potentials. The previous method, based on the exclusion of polarizing functions, though also effective, seems to be rather artificial, because the polarization terms have been estimated by purely mathematical manipulations. The procedure of applying hydrostatic pressure is similar to the so-called potential-induced breathing method (PIB) [130]. In contrast with PIB, however, we employ SCF-MO calculations in which covalency effects are taken into account.

Results concerning the octahedral $(NbO_6)^{7-}$ cluster are shown in Fig. 2.7. The satisfactory agreement between the empirical and the *ab initio* hydrostatic pressure potentials at relevant ion separations around 2 Å is remarkable. This result indicates that potential parameters derived by empirical fitting can have some physical relevance, thus giving support to shell model simulations based on careful empirical parametrizations. Both Nb^{5+}... O^{2-} potentials provide useful descriptions of $KNbO_3$ crystals (see Chap. 4). The deviations of the short-range potential which were derived from breathing mode cluster distortions are caused by the electronic polarization term $E_{\rm pol}$ discussed earlier. In the following we do not consider the latter potential. It can, however, be used in rigid ion potential models.

The differences between the NbO pair and the NbO_6 cluster results can mainly be traced back to complex covalency effects, which are largely suppressed in the present pair model. The ionic charges obtained by a Mulliken population analysis (see [17]) remain close to the pure ionic values in the

[5] It is noted that even at perfect lattice spacings there exists a certain asymmetry resulting from embedding the quantum cluster only within a field of point charges representing the outer crystal. However, cluster calculations with and without polarizing d-basis functions confirm that at perfect lattice spacings there is only a small polarization-induced contribution to the short-range interaction energy (≤ 0.1 eV) which may safely be neglected to first order. This becomes considerably different the more the niobium–oxygen separation deviates from the perfect lattice spacing.

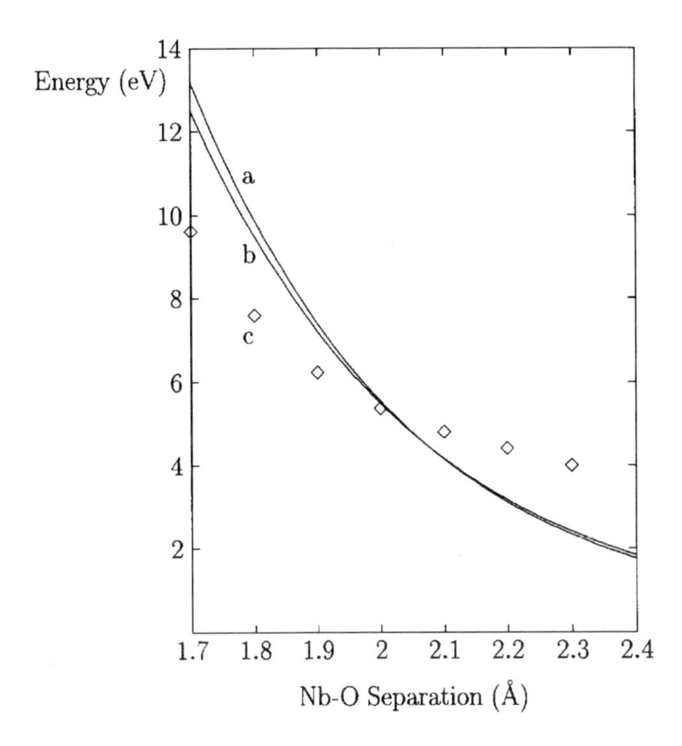

Fig. 2.7. *Ab initio* potential derived from an embedded NbO_6 quantum cluster. (a) Empirical Nb^{5+}... O^{2-} potential for comparison. (b) *Ab initio* hydrostatic-pressure Nb^{5+}... O^{2-} potential derived from a NbO_6 cluster. (c) *Ab initio* potential (◇) derived from breathing-mode distortions of NbO_6 within a fixed lattice

case of the pair model (Nb^{5+} : $q \simeq +4.75|e|$, O^{2-} : $q \simeq -1.75|e|$), but show significant deviations (particularly for the Nb ion) in the cluster model (Nb^{5+} : $q \simeq +2.84|e|$, O^{2-} : $q \simeq -1.64|e|$). In contrast with highly ionic systems, the various short-range interionic potentials in semi-ionic materials are not independent of each other. In $KNbO_3$ this concerns the Nb^{5+}... O^{2-} and the O^{2-}... O^{2-} interactions; by means of the NbO_6 cluster a consistent treatment of these potentials seems to be guaranteed. Pair cluster models, on the other hand, must obviously involve sophisticated embedding potentials (Sect. 2.1.2) to provide useful results. Because embedding potentials introduce additional short-range interactions with neighbouring crystal ions, the correspondingly improved pair cluster descriptions effectively go beyond the proper pair model. In contrast, the K^+... O^{2-} interaction is far more ionic in nature. Besides the O^{2-}... O^{2-} interaction we retain this potential from earlier simulation studies [131]. Table 2.1 summarizes all the short-range potentials which are used to model $KNbO_3$ (Chap. 4). Figure 2.8 shows a graphical display of these short-range potentials. The relative importance of the Nb^{5+}... O^{2-} short-range interaction is clearly seen.

Table 2.1. Short-range potential parameters for $KNbO_3$.

Interaction type	A (eV)	ϱ (Å)	C (eV Å^6)
O^{2-}... O^{2-}	22746.30	0.14900	27.88
Nb^{5+}... O^{2-} (*ab initio*)	1333.44	0.36404	—
Nb^{5+}... O^{2-} (empirical)	1796.30	0.34598	—
K^{+}... O^{2-}	1000.30	0.36198	—

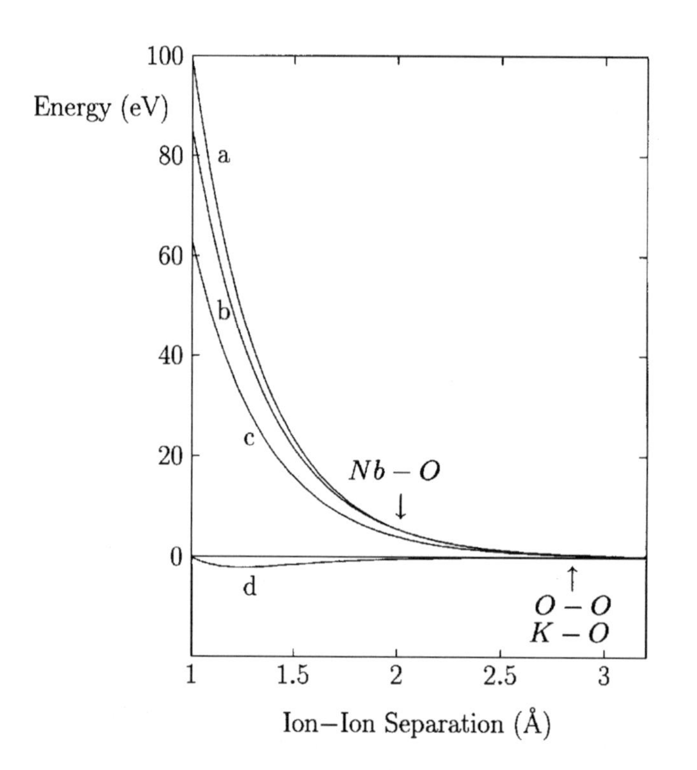

Fig. 2.8. Short-range potentials for $KNbO_3$. Arrows denote equilibrium ion–ion separations. (a) Empirical Nb^{5+}... O^{2-} potential. (b) *Ab initio* Nb^{5+}... O^{2-} hydrostatic potential from the NbO_6 cluster. (c) Empirical K^{+}... O^{2-} short-range potential. (d) Empirical O^{2-}... O^{2-} short-range potential

3. Barium Titanate

$BaTiO_3$ belongs to the family of perovskite-structured oxides. At room temperature it is stable in its ferroelectric tetragonal phase. The non-centrosymmetry of this phase forms a precondition for various photorefractive applications of $BaTiO_3$; similar remarks apply to other perovskite oxides, too. Section 3.1 reviews the most recent theoretical investigations towards an understanding of ferroelectricity. Figure 3.1 shows the unit cell for cubic $BaTiO_3$. This structural pattern is characteristic of all perovskite materials.

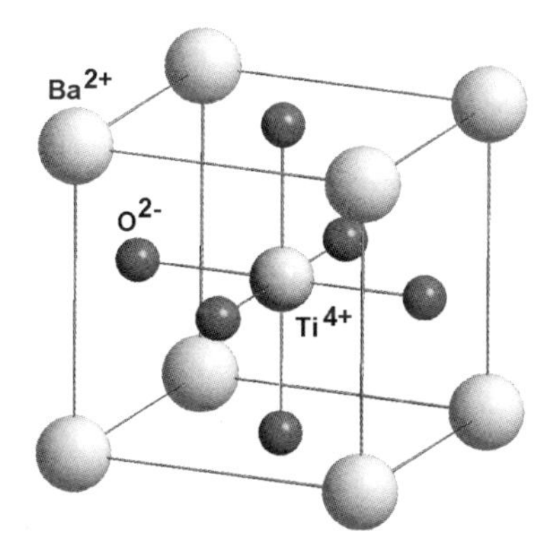

Fig. 3.1. The unit cell of cubic $BaTiO_3$

The performance of photorefractive materials is made possible by suitable point defects. Corresponding lattice imperfections are always generated during crystal growth. In particular, transition metal impurities (e.g. iron) play a leading role in this respect. The basic physics underlying the photorefractive effect may be summarized briefly as follows: as a result of interfering light beams striking a photorefractive oxide specimen, the crystal sample is exposed to light and dark regions. Defects in the light region, possessing electronic states in the band gap, are subject to light-induced charge transfer processes creating mobile charge carriers. Due to the Fermi level in as-grown $BaTiO_3$ these charge carriers are predominantly holes. Via diffusion, bulk photovoltaic effects and/or drift in pyroelectrical fields [132], the holes are transported into the dark regions where they are finally trapped in deep acceptor defect levels. However, before reaching their ultimate traps the holes can be intermediately bound in shallow energy levels which may be thermally ionized at room temperature. For example, alkali cations induce corresponding shallow band gap levels. Also, transition metal cations, being stable in three charge states, may induce shallow gap levels besides deep ones. Rhodium has been identified as falling into this category. Its stable charge states (i.e. +3, +4 and +5) introduce a corresponding two-level system [133]. It is noted that the interplay of deep and shallow defect levels leads to the observed sublinearity of the photoconductivity as a function of the light intensity (see [133] and references therein), i.e. $\sigma_{\mathrm{ph}} \propto I^{\alpha}$ with $\alpha < 1$. Finally, the space charge field generated in this way is transposed into a corresponding refractive index pattern by means of the electrooptic effect.

The major questions to be addressed in the present context concern the defect chemistry of $BaTiO_3$ and the electronic structure of relevant defects. The dominant defect structures in $BaTiO_3$ are extrinsic defects, of which transition metal cations and alkali cations provide important examples. Extrinsic defects may act as donor- or acceptor-type defects depending on their charge state and site of incorporation. Charge-compensating defects show a complementary donor/acceptor behaviour. Donors and acceptors are understood as electron and hole trapping defects, respectively. The actual importance of these defects depends on the particular position of the Fermi level. The defect chemical behaviour of $BaTiO_3$ is reviewed in Sect. 3.2. The corresponding investigations are based on shell model simulations.

Section 3.3 is devoted to electronic structure calculations in $BaTiO_3$. Considering defects, the transition metal cations (e.g. iron) are of particular interest. These impurity defects are able to generate deep levels in the band gap of $BaTiO_3$. In Sect. 3.3.1 we consider recent embedded cluster calculations on transition metal impurities in $BaTiO_3$, which have the advantage of including defect-induced lattice relaxations and electron correlations.

In Sect. 3.3.2 we discuss recent simulations of trapped holes. One may classify the possible trapping centres into acceptor defects either at Ti sites (e.g. transition metal impurities) or at Ba sites (as alkali cations). Particular attention will be paid to the localization properties of trapped holes. In Sect. 3.3.2 we also consider the formation of trapped hole bipolarons in $BaTiO_3$, which becomes possible if the actual trapping centre is able to bind at least two holes. Due to their diamagnetic ground state these bipolarons are invisible in traditional ESR experiments, but recent Photo-ESR investigations emphasize their role in light-induced charge transfer processes. It is interesting to note that hole bipolarons are also discussed in the context of oxide-based high-temperature superconductivity [8]. Moreover, hole bipolarons formed at the surface of La_2O_3 catalysts are believed to be responsible for methane activation [134].

Conduction band charge carrier transport becomes important upon shifting the Fermi level to higher energies. In this case the deep and shallow donor-type levels will strongly influence the material properties. Corresponding donor defects introducing shallow levels close to the conduction band are oxygen vacancies, which are predominantly formed as charge compensators of alkali cations. Their number reduces upon expelling the alkali ions from the lattice. In this case one can observe ESR signals due to isolated Ti^{3+} small polarons [133]. By analogy with holes one could also speculate on the formation of electron bipolarons. Preliminary calculations on these hypothetical bipolarons are reported in Sect. 3.3.3. It is noted that electron bipolarons have been established to exist in $LiNbO_3$ (see Chap. 5). Finally, in Sect. 3.3.4 I shall review the recent investigations on oxygen vacancies in $BaTiO_3$. It is emphasized that most of these latter results may safely be extrapolated to the other oxide perovskites.

3.1 Ferroelectricity of ABO_3 Perovskites

In this section a brief survey is given of recent theoretical efforts towards an understanding of ferroelectricity in perovskite-type ABO_3 oxides. The questions of major interest concern the ferroelectric driving forces, the dynamics of FE-PT (displacive versus order–disorder transitions) and the calculation of macroscopic polarizations. All subsequently mentioned investigations refer to potential simulations and *ab initio* descriptions of infinitely extended perfect crystals. For a discussion of finite size effects I refer to [135].

Accurate theories should account for the observed differences between the various perovskites. $BaTiO_3$ and $KNbO_3$ show the same sequence of FE-PT: *cubic* $\longrightarrow$ *tetragonal*, *tetragonal* $\longrightarrow$ *orthorhombic* and *orthorhombic* $\longrightarrow$ *rhombohedral*. Though being closely related to $KNbO_3$, $KTaO_3$ remains cubic for all temperatures. $PbTiO_3$ has only one FE-PT, i.e. *cubic* $\longrightarrow$ *tetragonal*. Unlike $BaTiO_3$, the related $SrTiO_3$ does not become ferroelectric.

Since Cochran's theory it has been commonly understood that ferroelectricity is closely related to a delicate balance between long-range Coulomb and short-range forces, of which the former favour ferroelectric displacements of ions whereas the short-range forces stabilize the cubic structure. The electronic structure of these materials plays an important role in this force balance. Shell model-based simulations, for example, emphasize the effect of oxygen polarizability. Corresponding calculations have been successfully employed to simulate phonon dispersion curves and the observed softening of TO phonons at the Γ-point [136, 137, 138, 139]. The effect of large anionic polarizabilities appears to be reasonable owing to the instability of the oxygen anions; the oxygen polarizability helps to make the effective interionic short-range potentials less repulsive. Potential-induced breathing (PIB) calculations for $KNbO_3$, taking additional non-spherical polarizations of oxygen anions into account, supported such an ionic interpretation [140].

However, the calculations emphasized a pronounced order–disorder character of the phase transitions. Generally, PIB model calculations [130] imply the superposition of in-crystal ionic charge densities to obtain the total electronic charge-density. Such investigations are thus in the spirit of an ionically bonded crystal.

On the other hand, the recent full-potential *ab initio* investigations suggest that the pronounced hybridization between B-cationic nd and oxygen 2p states is responsible for the occurrence of ferroelectricity (e.g. [141, 142]). For $BaTiO_3$ a comparison of a spherical ion PIB charge density with a full-potential LAPW (FLAPW) density has shown that the dominant density changes are localized close to the B cation sites [141] but not to the oxygen sites. Obviously electronic polarizability of ions as well as covalency effects can be responsible for ferroelectric distortions, but only the PIB model when extended to include ionic polarizations can unambiguously relate FE-PT to such polarizations. The situation is different when employing the shell model, because most shell model simulations are based on empirical parametrizations. The model parameters are highly dependent on each other; it is therefore not possible to unambiguously separate ionic polarizations from short-range interionic potentials (and thus from covalency). It cannot be excluded that large empirical polarizations are the consequence of pronounced hybridizations. A comparison of two different but closely related materials helps to clarify the interrelations. A recent shell model-based investigation of $KNbO_3$ and $KTaO_3$ [143] indicates that the B–O covalency, being modestly stronger in $KNbO_3$, reduces the effective anionic polarizability and increases that of the cation. The increased Nb polarizability destabilizes cubic $KNbO_3$ and, further, allows Nb dopants in $KTaO_3$ to move off-centre (see also Sect. 4.2.1).

The accuracy of all present *ab initio* investigations is limited by errors due to LDA, leading, for example, to a systematic underestimation of lattice constants by some hundredths of an Ångstrom. This appears to be crucial, because ferroelectricity is highly dependent on the volume. Further effects seem to be related to the specific band structure method being used. For example, FLAPW calculations [144] predict $KNbO_3$ to be cubic at a_{LDA}; pseudopotential [145] and FLMTO [48] investigations, on the other hand, yield ferroelectric distortions. Almost all calculations find the correct ferroelectric behaviour at the observed lattice spacings, which is then frequently taken for further investigations of the zone-centre potential surface. $KTaO_3$ represents an exception: FLMTO calculations [48] confirm this perovskite to be cubic at a_{LDA} but slightly ferroelectric at the observed lattice constant. These results suggest that $KTaO_3$ is close to an instability. Supposing that this instability is not due to computational inaccuracies, it could be possible that quantum fluctuations leading to some randomness which is not modelled so far would inhibit the calculated instability. This would support the idea of an incipient ferroelectric.

Usually the potential surface is analyzed with respect to phonon-related ion displacements (frozen phonon approach) and to strain variables. Dynamical properties related to the FE-PT can be studied by fitting the potential surface with suitable model potentials (e.g. [145, 146, 147]). Phonon frequencies can be calculated in this way, but it is also possible to use model Hamiltonians within a renormalization group theoretical approach to investigate the dynamical properties of ferroelectrics. This has been successfully done for the simple ferroelectric GeTe [148, 149] but not yet in the case of the more complex oxide perovskites. MD simulations have been done so far only within a pair potential approach. The corresponding potentials have been obtained from PIB calculations which were augmented by accurate FLAPW corrections [135].

Basic investigations of the potential surface show the existence of multiple well structures in the low-temperature phases, which is consistent with an order–disorder interpretation of FE-PT. The general interpretation emerging from the calculations essentially coincides with the eight-site model for B cations, which is assisted by the softening of TO phonons (e.g. [142]). However, it must be emphasized that all considerations are restricted to the centre of the first Brillouin zone, corresponding to a highly ordered structure. There is thus no account of the genuinely existing thermally stimulated randomness which may substantially affect the potential surface. Future studies must investigate these influences. *Ab initio* calculations further show the pronounced dependence of the potential surface (and thus of FE-PT) on volume and shear strains [142, 147, 145]. For example, a tetragonal strain $c/a \sim 1.06$ stabilizes the *ordered* tetragonal phase of $PbTiO_3$ over the rhombohedral phase, which is usually found as the ground state. This result explains the existence of only one FE-PT (*cubic* $\longrightarrow$ *tetragonal*) in this material. This FE-PT might be displacive in nature. In contrast to lead titanate, $PbZrO_3$ shows an antiferroelectric ground state. Recent LDA investigations [150] proved this structure to be slightly more favourable than a rhombohedral ferroelectric structure. The differences with respect to $PbTiO_3$ might be related to deviating hybridizations: whereas the Pb–O hybridization is weaker than in $PbTiO_3$, the Zr–O covalency is stronger than the Ti–O hybridization.

Pressure and strain effects may result from external influences but can also be of internal (chemical) origin. In the case of $PbTiO_3$ the lone pair 6s electrons of Pb^{2+} are largely responsible for the tetragonal strain. It seems that besides the degree of B–O hybridization the effect of lattice strains determines the observed differences between different oxides. A corresponding database may be found in [145]. Further, the proceedings of the second Williamsburg workshop on first-principles calculations for ferroelectrics provide a rich source on theoretical developments in this field (*Ferroelectrics*, vol. 136).

Finally, in agreement with experimental findings, recent FLMTO supercell calculations suggest the formation of chains of uniformly and along (cubic) [100]-axes displaced Nb ions in orthorhombic $KNbO_3$ [151]. These chains are perpendicularly orientated to the macroscopic polarization vector. Interactions between the chains favour the zero value of the average [100] Nb displacement.

The calculation of macroscopic polarizations and of dielectric material properties in polar materials could be based on density polarization functional theory (DPFT) (see Sect. 2.1). But, to my knowledge, this has not yet been attempted. Recently, a modern quantum theory of polarization has been developed by King-Smith, Vanderbilt, Resta et al. ([152, 153]). It is based on the observation that only polarization differences $\Delta \boldsymbol{P}$ between differently polarized states of one crystal are measurable bulk quantities. Theoretically, $\Delta \boldsymbol{P}$ is expressed as the integrated current flowing through the crystal upon polarization reversal. The electronic part of $\Delta \boldsymbol{P}$ has been identified as a Berry phase [154], i.e. as a gauge-invariant phase feature of the valence Bloch orbitals accompanying adiabatic changes round a circuit in parameter space. Resta et al. [155] and Zhong et al. [156] used this formulation to provide a first-principles approach to the spontaneous polarization of ferroelectric materials. Recently, Bernardini et al. [157] presented a polarization-based calculation of dielectric constants of polar crystals.

3.2 Defect Chemical Properties of Barium Titanate

Based on the empirical shell model potential parameter set derived by Lewis and Catlow [158] one may give the following overview of defect chemistry in $BaTiO_3$.

Intrinsic defects (of major importance are oxygen vacancies, but possibly also cation vacancies) are essentially formed only as charge-compensating partners of impurity cations. Purely intrinsic defect reactions, on the other hand, are not favourable. For example, the Frenkel defect formation energies range between 4.5 and 7.5 eV (per defect), and Schottky-like defects require 2.3–3 eV per created defect.

Vacancies may act as electron or hole traps. In qualitative agreement with empirical Green's function calculations of Fisenko et al. (see Sect. 3.3) shell model simulations suggest that oxygen and barium vacancies are comparatively shallow electron and hole trapping defects, respectively. Corresponding trapping energies have been estimated to be of the order of 0.1 eV. It is noted that within the shell model framework singly charged oxygen vacancies must be simulated as $V_O^{\bullet\bullet} - Ti_{Ti}^{l}$ defect complexes. However, this simple picture becomes significantly more complex within sophisticated embedded cluster calculations. In particular these investigations indicate the existence of certain deep gap levels which are absent in the shell model and in the Green's function simulations mentioned above. A detailed discussion of these recently performed investigations on oxygen vacancies is given in Sect. 3.3.4. Titanium vacancies, if they exist, define deep hole levels with trapping energies in the eV range.

Impurity cations dominate the defect chemical scenario in $BaTiO_3$. We first consider the incorporation of impurity cations with charge states ranging between +1 and +4. Generally, the precise incorporation energies (or chemical solution energies) largely depend on ion-size effects. As expected, the impurity size also determines the actual incorporation site. Impurity cations with charge states between +2 and +4 behave as donor defects when substituting for barium and as acceptor defects replacing titanium.

Favourable intrinsic charge-compensating defects are oxygen vacancies in the case of acceptor-type impurities and electrons for donor-type defects. The formation of compensating cation vacancies seems to be unfavourable.

Monovalent cations substitute for Ba cations with compensating oxygen vacancies. Small divalent impurity ions (e.g. Mg^{2+} and Ni^{2+}) replace Ti cations. Also these acceptor defects prefer the formation of oxygen vacancies. All remaining divalent impurities are incorporated at Ba sites without needing further charge compensation. The same is true for tetravalent cations, which substitute for titanium.

Of particular interest are trivalent impurities. Large cations can replace Ba cations, behaving then as donors. In this case electronic compensation is by about 6 eV more favourable than an alternative Ba vacancy compensation. Small trivalent cations like aluminium and most of the transition metal ions incorporate at Ti sites, which is again accompanied by formation of compensating oxygen vacancies. Besides these solution mechanisms, i.e.

$$M_2O_3 + 'BaTiO_3' \longrightarrow 2M^{\bullet}_{Ba} + 2e^{\prime} + 2BaO + \frac{1}{2}O_2(g) \tag{3.1}$$

$$M_2O_3 + 'BaTiO_3' \longrightarrow 2M^{\prime}_{Ti} + V^{\bullet\bullet}_{O} + 2TiO_2 \, , \tag{3.2}$$

self-compensation should be taken into account as a favourable solution mode:

$$M_2O_3 + 'BaTiO_3' \longrightarrow M^{\prime}_{Ti} + M^{\bullet}_{Ba} + BaTiO_3 \, . \tag{3.3}$$

Thus, trivalent cations can define donor- and acceptor-type defects at the same time. In $BaTiO_3$ self-compensation is principally restricted to trivalent cations[1]. Whether (3.1) may become preferred over self-compensation depends considerably on the electron formation energy in (3.1). By employing free-ion ionization potentials (3.1) may happen to be more favourable than (3.3) by no more than a few tenths eV. We shall return to this topic at the end of this section.

Extrinsic acceptor defects are able to bind oxygen vacancies. The precise binding energy depends on the impurity's charge and size misfit. In the case

[1] It has been argued [133] that self-compensation could impede the efficiency of photorefractive effects. However, this seems not to pose any problems in $BaTiO_3$ for two reasons: first, self-compensation is restricted to trivalent cations, and, second, almost all relevant transition metals avoid self-compensation in $BaTiO_3$. This situation will be different in $A^{1+}B^{5+}O_3$ oxides (see Chaps. 4 and 5).

of trivalent cations the binding energies typically range between 0.3 and 0.5 eV. They are only moderately smaller (i.e. 0.2 eV) for alkali impurities at Ba sites. It is interesting to note that the association of more than just one acceptor cation is likely: the binding energy of the second acceptor is almost precisely as large as the binding energy of the first acceptor cation. This outcome indicates that effective screening mechanisms of excessive charges in $BaTiO_3$ exist. In summary, shell model simulations suggest a pronounced association of (one or two) acceptor cations with the existing oxygen vacancies in $BaTiO_3$. Similar results may be expected for different oxide perovskites. This result seems to agree with the ESR-based observation that in almost all cases only tilted (i.e. non-axial) $V_O^{\bullet\bullet} - Ti_{Ti}^{l}$ complexes exist in reduced crystals containing alkali impurities [133]. The non-axiality of these centres refers to the associated acceptor cations at Ba-sites.

Finally, we remark on the incorporation of pentavalent cations (e.g. Nb^{5+}). The most favourable reactions involving cation vacancies or electrons read as:

$$M_2O_5 +' BaTiO_3' \longrightarrow 2M_{Ti}^{\bullet} + V_{Ba}^{ll} + BaTiO_3 + TiO_2 \tag{3.4}$$

$$M_2O_5 +' BaTiO_3' \longrightarrow 2M_{Ti}^{\bullet} + 2e^{l} + 2TiO_2 + \frac{1}{2}O_2(g)\,. \tag{3.5}$$

Upon using the fourth free-ion ionization energy for titanium the reduction solution mode (3.5) is by 2 eV more favourable than the competitive Ba vacancy mechanisms (3.4). But again we are faced with the problem that the accuracy with which electronic terms can be included significantly determines the precise energies of redox reactions.

At this stage, therefore, a brief digression is devoted to the calculation of crystalline electron affinities, which enter the estimation of reduction type energies within shell model simulations. Usually these affinities, which represent the formation energies of conduction band electrons e^{l}, employ free-ion ionization potentials corresponding to a perfect ionic crystal model. Thus, any covalent charge transfer contributions and crystal field effects are neglected in this way. Embedded cluster calculations may be used to estimate changes of ionization potentials upon going from free ions to crystal ions. For example, Hartree–Fock embedded cluster[2] calculations on TiO_6 clusters employing the perfect lattice geometry of $BaTiO_3$ yield a reduction of the fourth ionization potential of titanium corresponding to 1.4 eV. The analogous approach, when applied to embedded NbO_6 clusters, yields a reduction of 1.7 eV with respect to the fifth ionization potential of niobium. Electronic correlations which particularly increase the covalent charge transfer provide even

[2] The ECP parametrization of Hay and Wadt [21] has been used to simulate the central transition metal cation. The set of basis functions has been modified by completely breaking down the d-contractions. The oxygen ligand anions have been modelled by employing the split valence basis set of Dunning and Hay [126], which was further augmented with polarizing d-functions.

larger reductions. MP2 calculations (see Sect. 2.1) increase the Nb-related reduction to 3.0 eV, thus giving 47.55 eV for the in-crystal "fifth" ionization potential of niobium. In an analogous fashion 40.15 eV is obtained for the in-crystal "fourth" ionization potential for titanium

In particular, charge transfer directed from anions onto cations tends to destabilize electrons located at cation sites, whereas electrons at anions become stabilized. For titanium and niobium the resulting total level shifts are comparatively modest, because there is a partial cancellation of covalency-induced charge transfer contributions and the particular crystal field splitting terms for the cationic $d(t_{2g})$ levels. Similar effects are operative in the case of anions. Figure 3.2 vizualizes the corresponding energy shifts of levels. In most

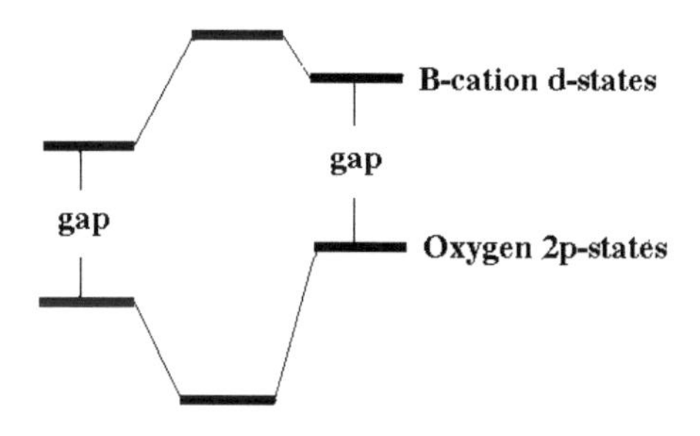

Fig. 3.2. Sketch of changes of B cation $d(t_{2g})$-levels and the upper O 2p-levels in ABO_3 perovskites due to covalent charge transfer (CT) and crystal field (CF) effects

simulation studies of ABO_3 oxides the band gap energies which were consistently calculated using free-ion ionization potentials and free-anion affinities provide satisfactory estimates of the observed band gaps of the particular materials (e.g., see [124, 158] and Chap. 4). This result might be mainly due to an almost parallel shift of cationic and anionic energy levels. However, in situations where either electron- or hole-formation energies enter the calculations (as in redox reactions) one may encounter significant deviations from the perfect ionic model.

In the case of $BaTiO_3$ there are a few defect chemical consequences when using the refined electron affinities: first, whereas smaller trivalent cations (e.g. Al^{3+}, Cr^{3+}, Mn^{3+} and Fe^{3+}) remain incorporated at titanium sites with compensating oxygen vacancies, self-compensation becomes the most favourable solution mode for all larger trivalent cations, thus leaving electronic compensation to be unfavourable unless there are electronic band gap levels (e.g. due to co-doping) effectively increasing the crystalline electron affinity. The possible exhaustion of co-doping induced gap levels would indicate the increasing importance of self-compensation with rising concentrations of trivalent rare earth cations. This behaviour seems to comply with observed dependences of resistivity on the donor content (see [158] and references therein). On may speculate that in all oxides co-doping effects attain general importance for low dopant concentrations. It is recalled, for example, that the defect chemistry in YIG is probably affected by co-doping phenomena (Sect. 7.5).

Second, in the case of pentavalent cations the electronic compensation reaction (3.5) remains only 0.2 eV more favourable than (3.4). However, it becomes slightly further stabilized by introduction of oxygen entropy contributions. Ba vacancies, on the other hand, could be formed only if high external oxygen pressures are applied.

3.3 Electronic Structure Calculations

Band structure calculations for perovskite-structured oxides have been performed extensively by Mattheiss using the augmented plane wave (APW) technique [63]. The calculations were based on an *ad hoc* crystal potential derived from Hartree–Fock–Slater atomic charge densities and on Slater's Xα method. The crystal potential was written as a Muffin Tin (MT) potential plus non-MT correction. The results suggest that the non-MT corrections are important to predict precise band structures. In $SrTiO_3$, being closely related to $BaTiO_3$, these corrections range from -1.1 to 0.3 Ry. The calculated band gap for $SrTiO_3$ is direct, but the author admits that this could easily be changed upon inclusion of self-consistency and by improving the exchange correlation terms. The calculated absolute gap value is about twice as large as the observed band gaps; this result originates from the omitted self-consistency and from the neglect of electronic correlations.

Recent state-of-the-art DFT-based band structure calculations (see e.g. [145]) are in qualitative agreement with the earlier band structures found by Mattheiss. However, these investigations also indicate an indirect band gap for these materials which is due to a valence band maximum at the X-point.

Michel-Calendini et al. [77] employed BO_6 cluster calculations based on the Xα and MT potential approximation in order to investigate the local electronic structure of ABO_3 perovskites (for limitations of the model see Sects. 2.1 and 5.1). Slater's transition state method has been used to calculate electronic excitation energies between the O 2p (VB) and B nd (CB) states determining the frequency-dependent electronic dielectric constant $\varepsilon(\omega)$ in $BaTiO_3$.

In a series of papers Michel-Calendini et al. (e.g. [159, 160, 161, 162]) reported their results of (spin-polarized) Xα cluster calculations for a number of transition metal impurities replacing Ti ions in $BaTiO_3$ or $SrTiO_3$. The investigations addressed the position of defect-related energy levels in the band gap, optical transitions and ESR parameters. The implementation of neutralizing Watson spheres implies complete screening of defect charges outside the MO_6 (or MO_5) clusters. On the other hand, all defect-induced lattice relaxations have been omitted from consideration. Depending on the size and charge misfits of impurities there would be substantial relaxations, particularly of the ligand anions, leading to pronounced shifts of the energy levels.

Selme and Pecheur [57, 58, 59] employed a Green's function description which was based on Mattheiss' tight binding parametrization of solid state matrix elements [63] in order to study defect levels in $SrTiO_3$ originating from oxygen vacancies and transition metal impurities (see also the paragraph on Green's functions in Sect. 2.1.2). The orbital description was restricted to include oxygen 2p and titanium 3d functions. Further, the defect-induced potential appearing in the Dyson equation (2.30) was assumed to be short-range, corresponding to complete screening beyond the first neighbours of the respective defect. As in the case of the Xα cluster calculations mentioned above, the effect of electronic screening is certainly overestimated in this way, whereas lattice relaxations have been essentially neglected. Although there might be a partial compensation of corresponding effects, the screening mechanism is probably too strong. In the case of transition metal impurities, spin-polarization of d-electrons has been introduced on the basis of the simplified expression $-\xi(n_{d\uparrow} - n_{d\downarrow})$, where the $n_{d\uparrow,\downarrow}$ are the numbers of spin-up and spin-down electrons and the Stoner constant $\xi \sim 1$ eV.

Fisenko and Prosandeyev et al. (e.g. [60, 61, 62]) performed further defect investigations in ABO_3 perovskites which were based on Green's functions. The treatment of the electronic structure of the undisturbed crystal is analogous to that of Selme and Pecheur. To model the spatial range of defect potentials the authors introduced an electronic screening sphere with complete screening outside this sphere. Defect levels related to oxygen and cation vacancies have been considered as a function of the vacancy charge state and the radius of the screening sphere. The vacancy defect potential has been defined to be infinitely strong; thus there is no allowance for any electron density transfer onto the vacant lattice site in this model (note that a similar approximation has been assumed in the investigations of Selme et al.). Donor levels were found for the oxgen vacancy and acceptor levels in the case of the cation vacancies. In particular, the B cation vacancies have been identified to define deep defect levels; oxygen vacancies, on the other hand, provide shallow electron levels close to CB. However, the results are certainly strongly dependent on the properties of the vacancy defect potential and on the screening sphere radius r_0. For r_0=a (lattice constant) the bound energy level of a singly charged V_O in $SrTiO_3$ corresponds to $\varepsilon_{CB} - \varepsilon = 0.04$ eV. (In spite of employing complete screening beyond nearest neighbours, Selme and Pecheur find the larger value of ~0.42 eV. The reason for this deviating result might be due to differences in the details of the perturbing vacancy potential.) In the limit of $r_0 \to \infty$, Fisenko et al. estimate 0.39 eV. These results emphasize the strong need for *ab initio* defect calculations (see Sect. 3.3.4) to avoid any effects of *ad hoc* assumptions.

The calculations of Fisenko et al. suggest that the bound levels of oxygen vacancies are predominantly localized on the two nearest Ti neighbours. Further investigations of Prosandeyev [61] employing an unrestricted Hartree–Fock description within a Green's function approach indicated a symmetry breaking of one-electron orbitals related to oxygen vacancies, i.e. electron localization can take place at exactly one Ti neighbour. For example, a singly charged oxygen vacancy should be interpreted as a V_O–Ti^{3+} complex[3]. Lattice relaxation may enhance this electron localization. Corresponding evidence derives from recent Green's function investigations of Prosandeyev et al. [62], taking simplified account of relaxation contributions. Thus, unlike proper ionic crystals, electrons at F centres in ABO_3 perovskites are attracted towards the nearest B cations, which is accomplished by covalency effects and stabilized by lattice relaxations. This interplay of covalency and lattice distortions seems to be typical for semi-ionic materials like the perovskite-structured oxides. All calculations convincingly prove that the B (or M) cation on which the electron resides moves away from the oxygen vacancy (see also Chap. 4).

I should emphasize at this stage that the symmetry-breaking or localization properties of the electron trapped at an oxygen vacancy depend sensitively on the detailed nature of the models discussed so far. For example, the neglect of excited Ti 4s and 4p orbitals as well as the implementation of the perturbing defect potential in the above Green's function studies should be viewed critically. This will be seen from a discussion of embedded cluster calculations on oxygen vacancies presented in Sect. 3.3.4.

The investigations of Michel-Calendini and Selme and Pecheur concerning the electronic defect properties of transition metal impurities in $BaTiO_3/SrTiO_3$ suggest that multiple charge states of a specified impurity species can easily be accommodated in perovskite oxides. However, the precise level positions depend to a considerable extent on the invoked complete screening assumptions. The predictions of defect level positions are therefore of a more qualitative value. In comparison to the Xα cluster calculations, the results of Selme and Pecheur generally underestimate spin polarization effects in favour of low-spin ground states. This depends on the simplified Stoner model involved. The Xα calculations are probably more reliable in this respect. In some cases the Xα investigations predict different ground states for $SrTiO_3$ and $BaTiO_3$. Because of the larger lattice parameter in $BaTiO_3$, high-spin states are preferred in this material. The calculations, however, neglect all defect-induced lattice relaxations which could modify the results.

[3] Inclusion of electron correlation would restore the full symmetry of the defect centre by appropriate superposition of the two equivalent symmetry-broken solutions. However, the time an electron spends near one of the Ti ions may be large on an experimental time-scale.

Ab initio embedded cluster calculations including defect-induced lattice relaxations (see Sect. 2.1.2) do not encompass artificial screening effects. They guarantee a self-consistent description of the perturbing defect potential and take lattice as well as electronic screening terms reasonably well into account. Details of corresponding calculations for transition metal impurities and hole-type defects will be discussed in the following subsections.

Recently, Cherry et al. [163] reported shell model and embedded cluster calculations (HF and MP2) on protons dissolved into oxide perovskites. In the embedded cluster studies only a subset of the cluster ions was allowed to relax from perfect lattice positions. Investigations for $SrTiO_3$ confirm that the OH^- dipoles formed are orientated along the edges of the oxygen octahedra, but orientations along $\langle 100 \rangle$ directions seem to be preferred in the $A^{3+}B^{3+}O_3$ systems studied. In particular, the MP2 calculations result in a very low energy barrier to proton transfer between neighbouring oxygen anions, indicating tunnelling processes in these oxides. However, the intermediate state preceding proton migration requires equivalent lattice environments for the neighbouring oxygens, and, according to shell model calculations, establishing this precondition needs some tenths eV.

The first *ab initio* calculations of $BaTiO_3$ surfaces have recently been reported by Padilla and Vanderbilt [164]. These authors considered BaO- and TiO_2-terminated (001) surfaces and investigated surface-induced relaxations and defect states, surface relaxation energies and the influence of the surface upon ferroelectricity: there are no deep gap surface levels, and the authors found only a small enhancement of ferroelectricity near the surface.

3.3.1 Embedded Cluster Calculations for Transition Metal Ions in $BaTiO_3$

As has been remarked in the introduction to this chapter, transition metal impurities sensitize the photorefractive effect in perovskite-structured oxides like $BaTiO_3$. The technological importance of these impurities is emphasized further upon noting that the development of tunable solid state lasers is based on transition metal cations doped into suitable host systems [165], e.g. Al_2O_3:Cr^{3+}.

In this subsection we mainly concentrate on discussing calculations for Mn^{4+} impurities solved into cubic $BaTiO_3$ at Ti sites. The reason for choosing Mn^{4+} as a test case is based on the observation that this impurity species represents a neutral defect when substituting for Ti^{4+}. Further, Mn^{4+} is isoelectronic to the important Cr^{3+}. Unfortunately, there is no experimental information for Mn^{4+} in cubic $BaTiO_3$. It has been argued [166, 167] that the pronounced dynamical order–disorder effects leading to ferroelectricity affect the Mn^{4+} and are responsible for the absence of data. In order to compensate for this deficiency the calculations are compared with observations of the isoelectronic $SrTiO_3$:Mn^{4+} instead. Both perovskite oxides are structurally very similar; however, unlike the barium compound, $SrTiO_3$ remains paraelectric at all temperatures.

The subsequent discussion is devoted to optical crystal field transitions (CF), defect-induced latice relaxations, Jahn–Teller distortions related to the excited $^4T_{2g}$ state and charge transfer transitions (CT). CT transitions are the dominant electronic excitations with respect to photorefractive effects.

Method. Since the present investigations aim at the accurate characterization of ground- and excited states, we shall mainly employ the Hartree–Fock approximation (if not explicitly stated we use the restricted open-shell HF formulation) augmented by suitable correlation treatments. Auxiliary DFT calculations, on the other hand, will be confined to investigations of charge transfer transitions. We recall that, due to employing ground state functional Kohn–Sham theory, the calculated energy separations between ground and excited states provide in most cases lower bounds to the exact energy differences (see Sect. 2.1). However, the quality of such lower bounds must in any instance be judged on the basis of different exact calculations.

The general procedure includes *ab initio* LCAO-MO calculations for a central $MO_6Ba_8Ti_6$ defect cluster, which simulates the crystal lattice immediately neighbouring an M_{Ti} transition metal impurity cation in $BaTiO_3$. The perfect crystal structure has been assumed according to the cubic high-temperature phase. The resulting 21 atom defect cluster $MO_6Ba_8Ti_6$ is visualized in Fig. 3.3. The size of the chosen quantum cluster appears to be sufficient for discussing the local electronic defect properties of interest. Employing larger clusters, on the other hand, might not be reasonable due to the increasing computational costs.

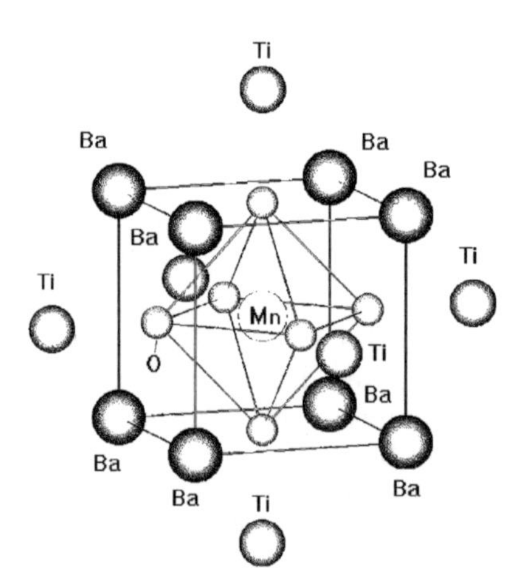

Fig. 3.3. The quantum mechanical defect cluster. The immediate environment of the central MnO_6 complex consists of eight barium and six titanium cations, which are represented by bare effective core potentials

To model the electronic structure of the cluster, effective core potentials (ECP) are employed in conjunction with double-zeta quality basis functions for the valence orbitals on manganese [21] and Dunning's contractions of Huzinaga's basis set for the oxygen ions [126]. The oxygen basis set is further augmented with diffuse p- and polarizing d- functions. The outer Ba and Ti ions are represented by bare effective core potentials which simulate ion-size effects of these cations. The *ab initio* ECP parametrization has been taken from Hay and Wadt [21].

Besides HF theory we shall mainly consider configuration interaction (CI) expansions to represent important electronic correlation contributions. These calculations are restricted to the inclusion of single and double electronic excitations (SDCI) with reference to the respective HF states. On the basis of perturbation theory the 15,000 energetically most important configurations will be chosen for a further diagonalization of the CI Hamiltonian matrix (the perturbation energy contribution of these kept states is denoted by PEK and the contribution of all neglected states by PEN). This procedure follows the earlier approach of Rawlings and Davidson [168]. Size consistency can be approximated using the formula of Davidson and Silver [169]

$$\begin{aligned} E_{\text{corr}}(\text{SDCI}+\text{Q}) = E_{\text{HF}} + \frac{c_{\text{HF}}^2}{2c_{\text{HF}}^2 - 1}\left[1 - \frac{\text{PEN}}{\text{PEK}}\right] \\ \times(E_{\text{corr}}(\text{SDCI}) - \text{E}_{\text{HF}})\,, \end{aligned} \tag{3.6}$$

where c_{HF} is the expansion coefficient of the Hartree–Fock wavefunction and $E_{\text{corr}}(\text{SDCI})$ the uncorrected SDCI energy. The stability of these SDCI(+Q) results is tested by calculating the natural orbitals, which by construction diagonalize the (SDCI) first-order reduced density matrix, and iterating the configuration interaction with these orbitals (NO-SDCI(+Q) procedure [170]). It is recalled that natural orbitals provide the most rapidly convergent CI expansion. In this sense the NO-SDCI(+Q) results are superior to SDCI(+Q) employing HF orbitals. Table 3.1 compiles energy separations between electronic crystal field states of Mn^{4+} which have been calculated employing these various CI approaches. In all cases SDCI+Q is superior to

Table 3.1. Calculated crystal field splittings employing the various types of CI as introduced in the text. The cluster geometry corresponds to cubic perfect lattice spacings

Type of CI	$\Delta_{\text{CF}} = E(^4\text{T}_{2\text{g}}) - E(^4\text{A}_{2\text{g}})/eV$
SDCI	1.15
SDCI+Q	2.48
NO-SDCI	2.11
NO-SDCI+Q	2.23

the uncorrected SDCI. The best results correspond to the NO-SDCI+Q level. As these calculations involve unreasonably large computer capacities, we may employ the following compromise: SDCI+Q will be used to determine the required total energy surfaces of the cluster, which are needed to perform the embedded cluster relaxation step (see below). Finally, energy separations are recalculated at the relaxed equilibrium configurations on the basis of NO-SDCI+Q. Therefore, if not explicitly specified all subsequently reported CI energy differences correspond to the NO-SDCI+Q level. The cluster *ab initio* calculations use the quantum-chemical program packages MELDF [171], HONDO 7.0 [172] and CADPAC [173], of which the latter code is employed to perform additional charge transfer calculations based on Møller–Plesset perturbation theory and DFT.

The embedding lattice is simulated by means of an interionic effective pair potential model, i.e. the shell model (see Sect. 2.2 for details). Ion charges are chosen correponding to a formal ionic model.

The short-range potential parameters appropriate for $BaTiO_3$ and for impurity–oxygen interactions are taken from the extensive work of Lewis and Catlow [158]. These pair potentials also specify the interactions of the cluster ions with the embedding shell model ions.

The total energy of the crystal is minimized with respect to the cluster (nuclear) coordinates R_c and to core and shell coordinates R_e of the embedding crystal ions. The additionally required minimization with respect to the cluster wavefunction Ψ describing the local electronic structure within the cluster region is performed by means of the *ab initio* HF(+CI)-SCF-MO calculations which were introduced above. The energy difference of the total energy of the quantum defect cluster $E_{QM}^{clus}(\Psi, R_c)$ corresponding to the electronic states Ψ of interest ($^4A_{2g}$ and $^4T_{2g}$ for Mn^{4+}) and its classical pair-potential counterpart $E_{SM}^{clus}(R_c)$ was calculated on a 5 × 5 × 5 mesh of a_{1g}-symmetrical breathing-mode displacements δ of O^{2-}, Ba^{2+} and Ti^{4+} ions surrounding the central impurity cation. These energy values, fitted to a fourth-order polynomial

$$\begin{aligned} P_\Psi(\delta_O, \delta_{Ba}, \delta_{Ti}) &= E_{QM}^{clus}(\Psi, R_c) - E_{SM}^{clus}(R_c) \\ &= \sum_{n+m+p \leq 4} A_{nmp} \delta_O^n \, \delta_{Ba}^m \, \delta_{Ti}^p \end{aligned} \tag{3.7}$$

in the three types of displacements, are used to update the total pair potential crystal energy and gradients as to include the embedding shell model as well as the quantum cluster contributions. The total energy of the crystal is given by (see Sect. 2.1.2, (2.57)):

$$E(\Psi, R_c, R_e) = E_{SM}^{crys}(R_c, R_e) + P_\Psi(\delta_O, \delta_{Ba}, \delta_{Ti}) \,, \tag{3.8}$$

with

$$R_c(i) = R_c^0(i) + \delta_i \;\; (i = \mathrm{O}, \mathrm{Ba}, \mathrm{Ti}). \tag{3.9}$$

R_c^0 denotes the unrelaxed positions of cluster ions; $E_{SM}^{crys}(R_c, R_e)$ is the shell model energy of the total crystal. The total crystal energy obtained by (3.8) comprises the substitution of a pair potential defect cluster by its quantum mechanical counterpart. The short-range cluster–lattice interaction is modelled on the basis of the known pair potentials.

The final crystal relaxation is performed using a modified version of the shell model program CASCADE [174, 175]. Modifications of CASCADE are indispensable, because the quantum cluster energy models an exact image of the *ab initio* breathing mode potential energy surface and, thus, goes beyond the pair potential approximation as used in CASCADE. It is sufficient to update only the energy and gradients appropriately, since CASCADE employs the variable metric technique [81] to minimize the crystal energy. Thus, the inverse Hessian is iteratively approximated using the updated coordinates and gradients and, finally, converges to its exact form.

Based on the cubic symmetry employed in the present problem, the quantum mechanical cluster configuration can be determined from *in vacuo* cluster calculations. This is true, since the first non-vanishing terms of the electrostatic crystal potential are of fourth order and, further, the cluster multipole moments vanish up to the hexadecapole moment. Generalizations to systems exhibiting lower than cubic symmetry would in principle be straightforward; however, the computational costs increase rapidly with the additional degrees of freedom. Moreover, the total quantum energy of the cluster may no longer be determined independently of the embedding lattice configuration. For related details the reader is referred to Sect. 2.1.2.

Lattice Relaxations. Table 3.2 displays the calculated breathing mode displacements of the three shells of ions which neighbour the central Mn impurity.

The manganese impurity is assumed in its electronic quartet states $^4A_{2g}$ and $^4T_{2g}$. Different degrees of approximations are employed to account for the local electronic structure of the defect cluster, i.e. descriptions based on the shell model, on the Hartree–Fock approximation and on configuration interaction. This order denotes the increasing flexibility of the modelled electronic structure. Pure shell model simulations are not able to reflect any changes which are related to the electronic state of the Mn ion (the corresponding columns in Table 3.2 are thus replicas of each other).

Inspecting Table 3.2, we observe that all results are in remarkable qualitative agreement: the pronounced inward relaxation of the Ba^{2+} ions, the outward relaxation of the Ti^{4+} ions and, finally, the inward relaxation of the oxygen ligands. The ligand relaxation is in agreement with ion size arguments, because $R(Mn^{4+}) < R(Ti^{4+})$ (see [176]). Its calculated size, however, depends on the degree of approximation applied to represent the electronic structure of the MnO_6 cluster. The more accurately it is described the less are the calculated ligand relaxations. Within the shell model the deficiencies of representing defect-induced electron redistributions are compensated

Table 3.2. Total ion relaxations within the first three neighbour shells of Mn^{4+}_{Ti} ($^4A_{2g}$ and $^4T_{2g}$ state). ECC denotes embedded cluster calculations. −: inward relaxation and +: outward relaxation

$^4A_{2g}$-related ion relaxation / Å			
Ion type	ECC(CI)	ECC(HF)	Shell model
O^{2-}	−0.035	−0.046	−0.240
Ba^{2+}	−0.347	−0.385	−0.190
Ti^{4+}	+0.113	+0.100	+0.01
$^4T_{2g}$-related ion relaxation / Å			
Ion type	ECC(CI)	ECC(HF)	Shell model
O^{2-}	−0.021	−0.037	−0.240
Ba^{2+}	−0.348	−0.370	−0.190
Ti^{4+}	+0.119	+0.106	+0.01

by exaggerated displacements of the oxygen ions. In comparison the oxygen displacements derived from embedded cluster calculations are smaller by an order of magnitude. Table 3.2 further shows that, in particular, the oxygen relaxations are more pronounced with the manganese cation being in its $^4A_{2g}$ ground state. Obviously, the excitation of the Mn^{4+} into the electronic $^4T_{2g}$ state slightly increases the manganese radius, since the excited state is more diffuse than the ground state. A similar argument may be applied to CI, which admixes extended excited states into the HF ground state, thus allowing for Coulomb correlations.

In order to distinguish between contributions due to a cluster–lattice mismatch and proper defect-induced relaxations we also consider the ion displacements for a perfect cluster containing Ti^{4+} instead of Mn^{4+} as the central cation. These calculations (Table 3.3) are performed within the Hartree–Fock approximation. CI is again expected to further reduce the oxygen ligand displacements. Two stages of a modelling are compared in Table 3.3. The first refers to a representation of the central Ti cation using a bare effective core potential (model a in Table 3.3). This description is fully symmetric with respect to the outer Ti cations at the cluster boundary. There are no covalency effects between cations and anions. In the second situation (model b) the electronic structure of the central Ti is treated equivalently to the Mn impurity; thus, only the 1s–2p core electrons are simulated by effective core potentials, whereas the outer electronic structure is given explicitly. In this case the symmetry between the central and the outer Ti cations is broken, since covalency and charge transfer can take place between the oxygen anions and the central cation. Obviously, these effects most importantly affect the radial relaxation of the Ba cations. Allowing charge transfer onto the cen-

Table 3.3. Total ion relaxations (Å) within the first three neighbour shells of a central Ti^{4+} cation. a: bare ECP for the central Ti, b: orbital representation analogous to Mn. −: inward relaxation and +: outward relaxation

Ion type	a	b
O^{2-}	+0.073	−0.030
Ba^{2+}	+0.022	−0.130
Ti^{4+}	−0.029	+0.021

tral Ti cation results in a pronounced inward displacement of the Ba ions. However, the overall compatibility between both cluster descriptions and the shell model representation of the outer lattice is remarkable, since even the Ba displacements in model b correspond to only 3.25% of the lattice constant.

Finally, the differences between the manganese and the titanium cluster (model b) may be interpreted as purely impurity induced: $\delta_{\mathrm{O}} = -0.016$ Å, $\delta_{\mathrm{Ba}} = -0.255$ Å and $\delta_{\mathrm{Ti}} = +0.079$ Å. These additional relaxations refer to the ${}^4\mathrm{A}_{2\mathrm{g}}$ ground state of Mn^{4+}. For the ${}^4\mathrm{T}_{2\mathrm{g}}$ state we obtain $\delta_{\mathrm{O}} = -0.007$ Å, $\delta_{\mathrm{Ba}} = -0.240$ Å and $\delta_{\mathrm{Ti}} = +0.085$ Å. The calculated defect-induced ligand relaxations (referring to ${}^4\mathrm{A}_{2\mathrm{g}}$) are in satisfactory agreement with reported effective ion size differences (0.065 Å) [176]. Generally, the defect-induced relaxations will increase with increasing charge and size misfit of the impurity cation. Indeed, preliminary HF simulations for $\mathrm{Cr}_{\mathrm{Ti}}^{3+}$ employing a smaller $3 \times 3 \times 3$ mesh of $\mathrm{a}_{1\mathrm{g}}$-type displacements yield the following defect-induced relaxations: $\delta_{\mathrm{O}} = +0.068$ Å, $\delta_{\mathrm{Ba}} = -0.27$ Å and $\delta_{\mathrm{Ti}} = +0.12$ Å for the ${}^4\mathrm{A}_{2\mathrm{g}}$ state and $\delta_{\mathrm{O}} = +0.072$ Å, $\delta_{\mathrm{Ba}} = -0.22$ Å and $\delta_{\mathrm{Ti}} = +0.11$ Å for the ${}^4\mathrm{T}_{2\mathrm{g}}$. In particular, the outward displacements of the oxygen ions are larger by an order of magnitude than the reported radii differences between Ti^{4+} and Cr^{3+} [176]. This result suggests the dominating effects due to charge misfit and confirms earlier shell model-based simulations of Sangster [177].

In the following subsections we prefer to use the total embedded cluster relaxations (see Table 3.2) over the purely defect-induced displacement contributions, because only the first represent fully *equilibrated* lattice structures.

Optical Absorption and Emission between Crystal Field States. Table 3.4 displays the calculated energy separations between the excited ${}^4\mathrm{T}_{2\mathrm{g}}$ and ${}^2\mathrm{E}_{\mathrm{g}}$ states and the electronic ground state ${}^4\mathrm{A}_{2\mathrm{g}}$. These calculations employ the ΔSCF method which takes all important orbital relaxation effects into account (see also the paragraph on stability contained in this subsection and Sect. 2.1.2). Three different lattice geometries are considered, i.e. observed perfect lattice spacings and $\mathrm{a}_{1\mathrm{g}}$-relaxed lattices corresponding to the ${}^4\mathrm{A}_{2\mathrm{g}}$ and ${}^4\mathrm{T}_{2\mathrm{g}}$ electronic states of Mn^{4+} (see the paragraph on lattice relaxations in this subsection, Table 3.2). The different lattice geometries obtained from HF and CI are also taken into account. However, at this stage we neglect further JT distortions, which are to be expected for the orbitally

Table 3.4. Energy separation between crystal field states. o A: optical absorption and o E: optical emission

Electronic state	HF energy separation (eV) from $^4A_{2g}$ for different lattice geometries		
	undistorted	$^4A_{2g}$ relaxed	$^4T_{2g}$ relaxed
^2E_g	1.87	1.84	1.84
$^4T_{2g}$	2.31	2.56 (o A)	2.53 (o E)
Electronic state	**CI energy separation (eV) from $^4A_{2g}$ for different lattice geometries**		
	undistorted	$^4A_{2g}$ relaxed	$^4T_{2g}$ relaxed
^2E_g	1.65	1.65	–
$^4T_{2g}$	2.23	2.40 (o A)	2.37 (o E)

degenerated excited states. The transition energies calculated with respect to the $^4A_{2g}$ and $^4T_{2g}$ equilibrated lattices correspond to optical absorption and emission, respectively. HF derived excited state energy separations from the ground state are 0.1–0.2 eV greater than the corresponding CI energy differences. Besides a direct energetic effect, which may be inferred by comparing HF and CI energy differenes calculated in the *same* lattice structure (for example, compare the perfect lattice results in Table 3.4), there is an indirect effect which is based on the differences in the relaxation patterns produced by HF and CI (see the paragraph on lattice relaxations, Table 3.2). In the case of CI the smaller inward displacements of the oxygen ions lead to a reduction of the crystal field splitting $E(^4T_{2g})–E(^4A_{2g})$. For example, taking the two $^4A_{2g}$ equilibrated geometries of Table 3.2 we obtain a reduction of 0.06 eV. Generally, the effect turns out to be $\leq$0.1 eV.

The calculated Stokes shift, i.e. the difference between absorption and emission energies, is small and equals 0.03 eV within HF and CI theory. The same order of magnitude has been found for Cr^{3+} doped into MgO [178]. Because there are no experimental data for Mn^{4+} in cubic $BaTiO_3$, we compare the calculations with measurements on $SrTiO_3$:Mn^{4+}: $E(^2E_g)–E(^4A_{2g})$=1.71 eV and $E(^4T_{2g})–E(^4A_{2g})$=2.14–2.23 eV (absorption) [179, 180, 181]. Based on the smaller lattice constant in this material there will be essentially no ligand relaxation around Mn^{4+} in $SrTiO_3$; thus the experimental energies for $SrTiO_3$:Mn^{4+} may safely be compared with calculated energies in $BaTiO_3$ which employ relaxed lattices. Whereas the ^2E_g-related separation is insensitive to lattice relaxations, there is a pronounced effect for $E(^4T_{2g})–E(^4A_{2g})$. The calculated absorption energies (CI) are by ~0.2 eV greater than the reported experimental data. The relaxation contributions due to a cluster–lattice mismatch (see the paragraph on lattice relaxations in this subsection) could be mainly responsible for this deviation. In spite of this feature the agreement between calculated and experimental energy separations is very encouraging. However, the present discussion also suggests the importance of including electron correlations in order to predict accurate crystal field splittings.

The orbital degeneracy of the excited $^4T_{2g}$ state leads, under the action of appropriate electron–phonon couplings, to the occurrence of symmetry-reducing Jahn–Teller distortions. These JT distortions are approximately treated by minimizing the total energy with respect to the symmetry-adapted displacements of the oxygen ligands with the remaining ions fixed in their a_{1g}-relaxed positions corresponding to the $^4T_{2g}$ state. Figure 3.4 displays the symmetry adapted normal modes of an octahedron.

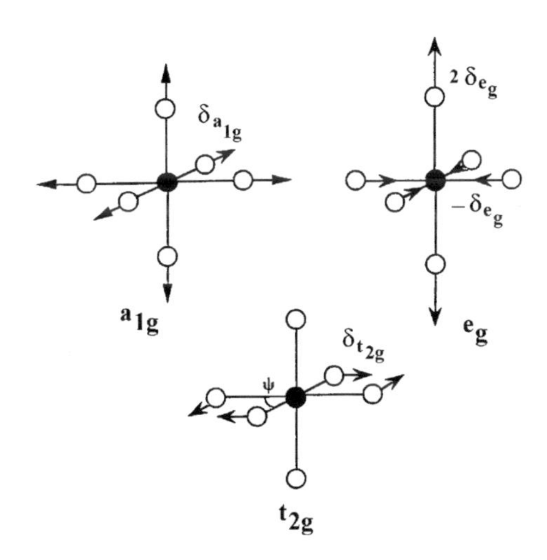

Fig. 3.4. Symmetry-adapted a_{1g}, e_g and t_{2g} distortions of an MO_6 octahedron

The e_g–$^4T_{2g}$ and t_{2g}–$^4T_{2g}$ electron–phonon interactions are JT active. The present HF and CI calculations do not show any significant t_{2g}–$^4T_{2g}$ instability; therefore in what follows we may concentrate on the e_g mode coupling. Jahn–Teller energies and frequencies are obtained by fitting the corresponding potential energy curves (PEC) with polynomials defined up to sixth order in the JT-mode displacements. Whereas in the case of HF-PEC we obtain good quality fits (corresponding to mean deviations <0.01 eV) even in the harmonic approximation (i.e. employing parabolas), it is necessary to consider sixth order terms for CI-PEC in order to maintain the quality of our fits. This result suggests that correlation effects increase the anharmonic potential terms. Table 3.5 compiles calculations of e_g mode couplings. All results are based on the best fits employing sixth-order polynomials. The frequencies are obtained from Taylor expansions to second order around the respective curve minima. Inspection of Table 3.5 shows that the introduction of correlation lowers E_{JT} and increases the vibration frequencies. An analogous influence on frequencies can also be observed in the case of the breathing mode a_{1g}–$^4A_{2g}$ electron–phonon couplings (HF: 830 cm^{-1} and CI: 980 cm^{-1}). Table 3.6 displays calculated ion charges obtained from a Mulliken population analysis (MPA) for the excited $^4T_{2g}$ state. These results are based on HF and NO-SDCI charge densities which have been calculated employing the a_{1g}-relaxed $^4T_{2g}$(CI) lattice. The occupied Mn^{4+} e_g orbital corresponds to $d_{x^2-y^2}$.

Table 3.5. Coupling of local e_g modes to $^4A_{2g}$and $^4T_{2g}$. We note that $\delta(O_{\pm z})=2\delta_{e_g}$and $\delta(O_{xy})=-\delta_{e_g}$

	HF		CI	
	$^4A_{2g}$	$^4T_{2g}$	$^4A_{2g}$	$^4T_{2g}$
δ_{e_g} / a.u.	–	−0.029	–	−0.017
E_{JT} / cm^{-1}	–	−290	–	−138
ω_{JT} / cm^{-1}	+650	+652	+768	+710

Table 3.6. Ion charges calculated from an MPA based on the excited Mn^{4+} $^4T_{2g}$ state. Further details are given in the text

Ion	Charge (HF)	Charge (NO-SDCI)
Mn	+2.37	+2.23
O (xy-plane)	−1.75	−1.72
O ($\pm z$)	−1.68	−1.67

We emphasize the relative merit of MPA charges, according to which only the differences between the HF and CI analyses are physically significant but not the calculated ion charges (e.g. see [17] for details). It can be seen that CI increases the charge transfer onto the central manganese cation, which results in an enhanced bonding stiffness and, thus, leads to the prediction of higher JT frequencies; at the same time correlation reduces the the charge differences between the xy-planar oxygen ions and the ones along $\pm z$, which is in line with a reduction of the JT distortion δ_{e_g}. Figure 3.5 shows the calculated $^4T_{2g}$ total HF and CI energies as a function of an e_g-JT distortion. Figure 3.6 displays a configuration diagram related to the e_g mode.

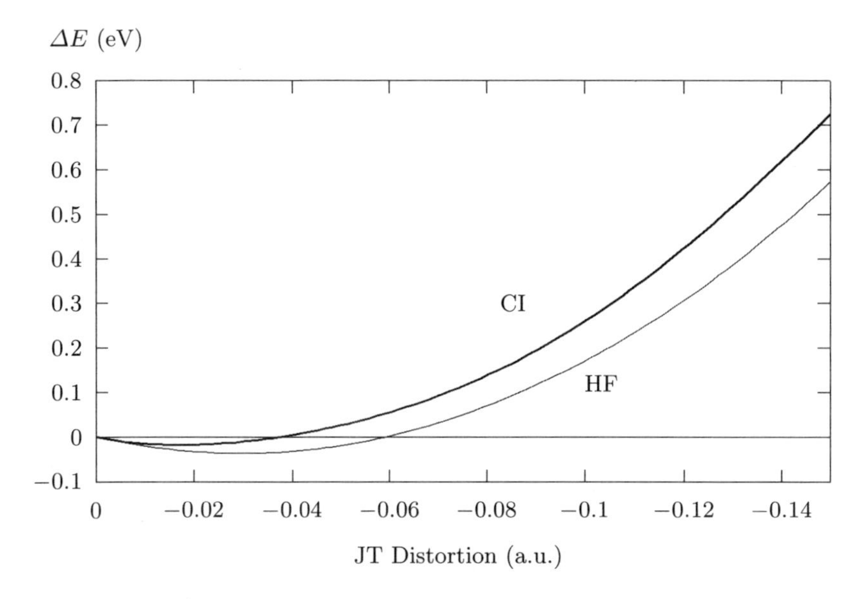

Fig. 3.5. e_g–$^4T_{2g}$ Jahn–Teller effect. HF and CI potential energy curves are shown as a function of δ_{e_g}

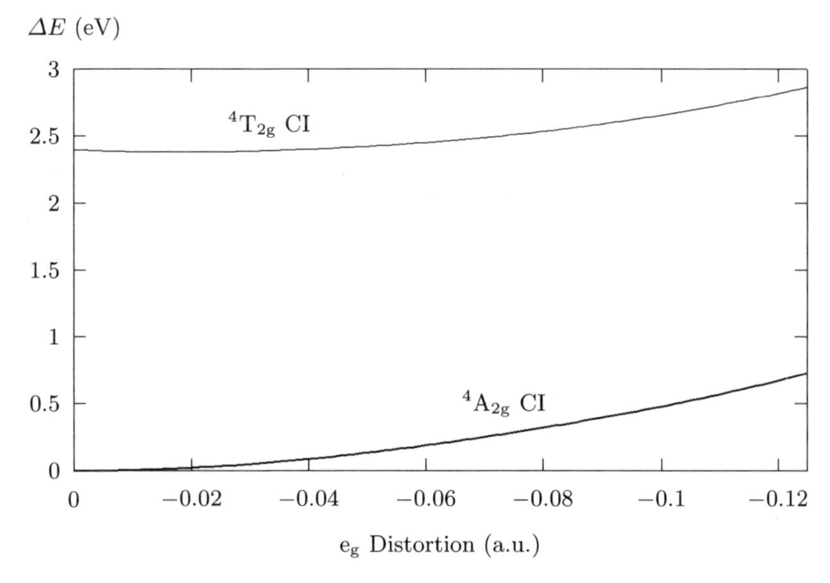

Fig. 3.6. Configuration diagram displaying the energy dependence of the $^4A_{2g}$ and $^4T_{2g}$ states upon e_g-type ligand distortions δ_{e_g}

Finally, the present simulations of JT distortions suggest a pronounced dynamical behaviour, since the JT energies are small in comparison with the frequencies.

Stability and Charge Transfer Transitions. In this subsection we comment on the relative stability of Mn_{Ti}^{4+} defect centres in $BaTiO_3$.

As has been demonstrated in Sect. 2.1.2, there are two main reasons why physical interpretations of embedded cluster calculations should not be based on cluster orbital energies. First, cluster eigenvalues correspond to localized crystal orbitals and are thus not directly related to the crystalline eigenstates which provide information on defect levels in the band gap. Second, orbital relaxation effects generally prevent a useful interpretation of cluster eigenvalues as ionization energies. It is recalled from Sect. 2.1.2 that Mn_{Ti}^{4+} would be unstable if we trusted the the misleading orbital-related information.

In order to establish the relative stability of the Mn_{Ti}^{4+} centres it is necessary to investigate charge transfer transitions (CT) involving an electron transfer from the oxygen ligands onto the manganese cation. The present CT ΔSCF calculations are performed employing perfect lattice spacings as well as relaxed lattice structures. For the latter type of calculation we adopt a further approximative relaxation procedure in order to account for symmetry-reducing lattice distortions accompanying the CT states. This method goes beyond the treatment of JT distortions as outlined in the discussion of optical absortions and emissions between crystal field states (this subsection), since it allows the symmetry-reducing lattice displacements to occur for all cluster ions and not only for the ligand anions. Such a generalization seems to be necessary, because the modifications of the oxygen electron structure due to CT are obviously more pronounced than in the case of the $^4T_{2g}$ CF state. The outer crystal lattice represented by 736 point ions with integral charges has been held fixed, corresponding to the relaxed crystal structure, which is in equilibrium with the Mn^{4+} $^4A_{2g}$ HF ground state. The cluster relaxation is performed using the geometry optimization facilities of the HONDO 7.0 *ab initio* code. For this task the program has been modified in order to include short-range Buckingham potentials between cluster ions and embedding point charges. Table 3.7 lists the corresponding total energy differences (see (3.8)) between the CT states and the Mn^{4+} $^4A_{2g}$ state. The CT states invoke the formation of Mn^{3+} for which we assumed two possible configurations, i.e. the high-spin configuration $(t_{2g})^3e_g$ (^5E_g) and the low-spin configuration $(t_{2g})^4$ ($^3T_{1g}$). The columns a and b in Table 3.7 refer to these situations, respectively. The difference between these energies measures the energy separation between the two Mn^{3+} crystal field states.

The Mn^{3+} high-spin configuration is by 1.8–1.9 eV more favourable than the low-spin configuration. In the case of the high-spin configuration of Mn^{3+} the hole created localizes on the two oxygen anions along $\pm z$, because the occupied Mn^{3+} e_g orbital corresponds to $3z^2 - r^2$. In this case the electronic structure of the acceptor type cation (Mn^{3+}) determines the localization

Table 3.7. Energy separation (eV) between CT states and the Mn^{4+} ground state ${}^4A_{2g}$, $\Delta E = E(CT) - E({}^4A_{2g})$. Columns a and b refer to the high-spin configuration $(t_{2g})^3 e_g$ and to the low-spin configuration $(t_{2g})^4$ of Mn^{3+}, respectively. The energies are obtained from UHF calculations, whereas the numbers in brackets refer to ROHF. In the case of relaxed lattices each electronic state is calculated within its own equilibrium lattice geometry

Lattice	a	b
perfect	1.40 (1.13)	3.21
relaxed	0.77 (0.5)	2.66

properties of the associated hole. More generally, it would be important to investigate the hole localization as a function of size, charge state and electronic structure of the trapping acceptor defect. For example, in suitable cases the trapping of two holes may be more favourable than the association of only one hole, which can inititate the formation of hole bipolarons (see Sect. 3.3.2).

Table 3.7 shows that Mn^{4+} is stable against CT, since the ${}^4A_{2g}$ ground state of Mn^{4+} represents the most favourable electronic state irrespective of the inclusion of lattice relaxation. However, lattice relaxations lead to a reduction of the relevant energy separations. This is reasonable, because Mn^{4+} is almost identical to the substituted Ti^{4+}; the CT state, on the other hand, defines a comparatively larger perturbation of the local crystal structure which needs stronger lattice relaxations for its compensation or screening. The additional influence of electronic correlations may be investigated on the basis of Møller–Plesset perturbation theory to second order (MP2) and DFT. Qualitatively, the effect of correlations may be understood on the basis of increased covalency between the oxygen anions and the manganese cation. In comparison to HF, electrons on the anions become more stabilized but destabilized on the cation. As a consequence the CT states are shifted to higher energies. Preliminary MP2 calculations, which employ the previous basis set (see the paragraph on methods contained in this subsection) as well as perfect lattice spacings, indicate a corresponding energy shift of about 0.6 eV (see also below).

Finally, we consider the energy separations between the Mn^{4+} ${}^4A_{2g}$ state and the lowest CT state involving for both states consistently either the ${}^4A_{2g}$ relaxed lattice or the CT relaxed lattice. In the first situation the energy separation corresponds to optical absorption and in the second case to optical emission (see Fig. 3.7).

Within the UHF approximation 1.1 eV is obtained as the onset value for the CT optical absorption and 0.4 eV correspondingly for the CT optical emission. First, we observe that CT transitions are characterized by large Stokes shifts. As expected, this is different from the crystal field transitions discussed above. Second, however, the calculated onset energy of the CT op-

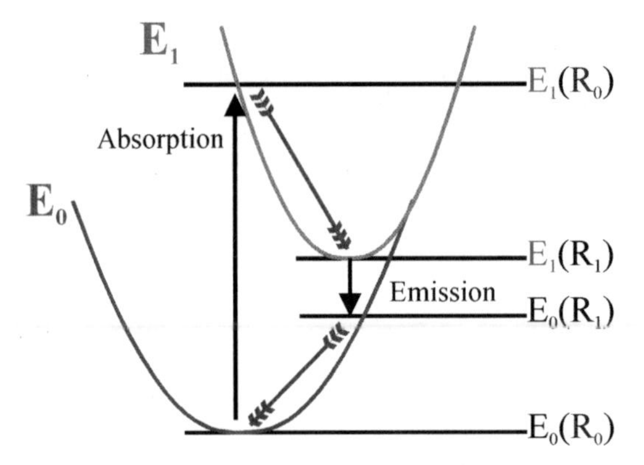

Fig. 3.7. Schematic visualization of optical absorption and emission. The vertical arrows denote electronic CT transitions which, according to the Franck–Condon principle, take place at fixed ion core positions. The displacement of the excited state potential energy surface (PES) against its ground state counterpart measures the Franck–Condon shift. Assuming equal PESs the Franck–Condon shift may also be characterized by the energy difference $E_1(R_0) - E_1(R_1)$

tical absorption is somewhat smaller than the reported experimental energies (> 3.2 eV) referring to $SrTiO_3$:Mn_{Ti}^{4+} [179]. In particular, the CT absorption energy is significantly smaller than the $^4T_{2g}$–$^4A_{2g}$ transition energy, which contradicts the experiments. Because the predicted energy corresponds to the HF approximation, it can be suggested that the inclusion of electron correlations is necessary and sufficient to calculate accurate CT transition energies.

MP2 calculations based on the previous basis set (see the paragraph on methods in this subsection) and employing the $^4A_{2g}$ equilibrated crystal lattice yield 2.0 eV for the CT optical absorption. This value represents a lower bound to the true CT energy, because in contrast with the crystal field splitting energies the CT transition energies are found to be highly dependent on the quality of the used basis set. The reason for this different behaviour is based on the pronounced charge redistributions accompanying CT. The accurate simulation of CT, therefore, needs a very flexible basis set. For example, using a full-orbital split valence basis set for the manganese cation, retaining the previous description of the oxygen anions (see the paragraph on methods in this subsection) and, finally, augmenting this basis set by additional oxygen and manganese d-type functions increases the UHF and MP2 CT absorption energies to 1.7 eV and 3.3 eV, respectively. It is noticed that even with this sophisticated basis set the UHF CT absorption energy is significantly smaller than the $^4T_{2g}$–$^4A_{2g}$ CF transition energy, which contradicts to all relevant experiences. But note that in this case MP2 predicts a reasonable CT absorption energy, which is significantly greater than the $^4T_{2g}$–$^4A_{2g}$ CF splitting and which agrees satisfactorily with reported experimental data for $SrTiO_3$:Mn_{Ti}^{4+} [179].

Further simulations can be performed on the basis of density functional theory (DFT). Following the investigations of Perdew and Levy [39] (see also Sect. 2.1) the $^4A_{2g}$, $^4T_{2g}$ and CT electronic state energies are calculated employing the $^4A_{2g}$-state equilibrated lattice geometry, which was derived earlier within HF theory. The present calculations, based on the Kohn–Sham procedure implemented in the CADPAC code [173], employ the same sophisticated basis set as the MP2 calculations discussed above. The exchange correlation functional is approximated by the advanced "BLYP" functional (Sect. 2.1).

The results of the present DFT simulations are encouraging: whereas the $^4T_{2g}$ is 2.1 eV above the $^4A_{2g}$ ground state, the CT absorption energy becomes 3.4 eV. These energy separations, which are expected to give lower bounds to the true excited state separations from the ground state, are in good agreement with the corresponding values obtained from the SDCI+Q and MP2 calculations discussed above. Most importantly we notice that the CT absorption energy is at least 1.2 eV greater than the respective CF transition energy. Experimentally, for $SrTiO_3{:}Mn^{4+}$ a corresponding energy separation of $\sim$1 eV has been observed [179].

Results of similar quality are obtained, when studying optical absorption CT transitions in the case of Fe^{4+}_{Ti}, i.e. $Fe^{4+}_{Ti} + h\nu \longrightarrow Fe^{3+}_{Ti} + O^-_{del}$. O^-_{del} denotes a delocalized hole state (see Sect. 3.3.2). DFT calculations yield an absorption energy of 2.3 eV, which is only slightly less than the reported experimental value ($\leq$ 2.8 eV [133]). In this context I add a remark on rhodium-related CT transitions. It has been proposed [133] that the CT process $Rh^{4+}_{Ti} + h\nu \longrightarrow Rh^{3+}_{Ti} + O^-$ shows a significantly smaller Franck–Condon shift compared with Fe. A simplified explanation for this behaviour may be given by reference to Fig. 3.7. $E_1(R_0) - E_1(R_1)$ is assumed to measure the Franck–Condon shift. Whereas iron represents an intermediate crystal field defect preferring a high-spin configuration, rhodium belongs, due to its 4d-electrons, in the strong-field case, and favours low-spin configurations. As a consequence, the excited CT electron will occupy an e_g orbital in the case of iron, but a t_{2g} state referring to rhodium. The electron–lattice coupling is stronger for e_g orbitals, since these point directly to the octahedrally coordinated oxygen ligands (see Fig. 3.8). Indeed, embedded cluster calculations indicate that the the CT-induced ligand relaxations are twice as large for iron as for rhodium. Therefore the iron-related energy difference $E_1(R_0) - E_1(R_1)$ is also larger. Preliminary investigations (MP2) yield 1.4 eV for iron, but only 0.4 eV in the case of rhodium.

The remarkable quality of the present DFT results may at least partly be traced back to significant electron density differences between the requested states. However, it is recalled in this context that there are also situations where DFT fails to give reliable excited state estimates. Most prominent is the inability of the usual DFT calculations to reproduce the atomic multiplet structure [39].

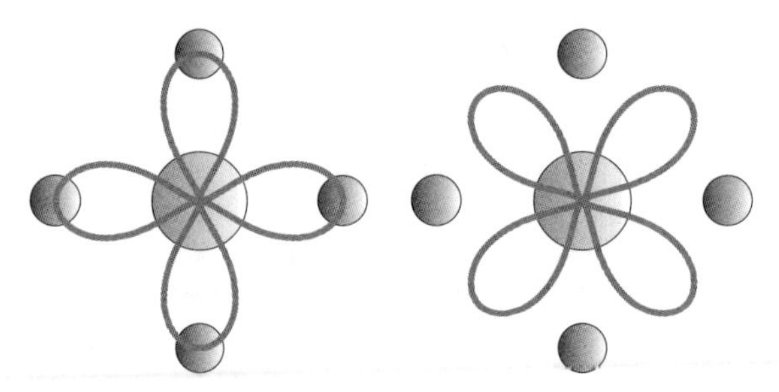

Fig. 3.8. Visualization of $e_g(3d_{x^2-y^2})$ (left figure) and $t_{2g}(3d_{xy})$ (right figure) orbitals centred on a transition metal cation. Oxygen ligands are simulated by shaded spheres

In conclusion we observe that electron correlation contributions are necessary and sufficient in order to reliably simulate CT transitions related to transition metal impurities in $BaTiO_3$. The suggested embedded cluster approach gives useful CT transition energies provided that the employed set of basis functions is sufficiently flexible. In this context we also briefly comment on the occurrence of symmetry-broken CT states within HF theory, which simulate a complete hole localization at exactly one of the oxygen ligand anions. These localized CT states, though representing the most favourable HF states, do not indicate the instability of particular impurity charge states in $BaTiO_3$, but refer to a general instability due to HF theory which is a manifestation of the so-called symmetry dilemma [182, 183]. On the other hand, the symmetrized CT states discussed so far reflect the stability properties of impurity cations in $BaTiO_3$ at least qualitatively. In fact, after inclusion of sufficient electron correlations the symmetrized CT solutions represent the most favourable CT states for all investigated transition metal cations. Based on this observation we have discarded the localized CT states from the present considerations in order to avoid these additional HF specific problems. In Sect. 3.3.2 we shall return to this topic.

Moretti et al. [91] reported a calculated CT transition energy of $\sim$2.8 eV. These cluster calculations were based on the $X\alpha$ exchange approximation and on perfect lattice spacings. Further, these investigations neglected any correlation contributions. The reason, that this value is about 1 eV greater than the most accurate UHF-based CT energy threshold derived from the present embedded cluster calculations, is not completely clear. However, it could be possible that the local $X\alpha$ exchange potential underestimates the gain in exchange energy between t_{2g}- and e_g-type electrons which occurs upon forming the Mn^{3+}. Moreover, certain shortcomings could also be due to the employed Muffin Tin potential approximation.

3.3.2 Simulation of Trapped Holes

Light-induced charge transfer reactions can be investigated on a microscopic level by means of photo-electron spin resonance techniques (Photo-ESR), i.e. by detection of ESR changes as a function of the irradiated wavelength. Corresponding Photo-ESR experiments [184] furnished evidence that trapped holes are significantly involved in light-induced charge transfer processes in $BaTiO_3$. These hole centres are either para- or diamagnetic. There are deep and shallow band gap levels of trapped holes. In most cases, the deep levels belong to transition metal impurities. Due to their ESR characteristics the paramagnetic holes referring to shallow levels may be identified with O_O^--type centres trapped at suitable acceptor imperfections. The diamagnetic species, on the other hand, are only indirectly accessible to Photo-ESR experiments, see Fig. 3.9 (reproduced from [184]). The figure displays a correlation between the observed ESR intensities of paramagnetic defects and the wavelength Λ of the illuminating light. Holes created in the valence band by photo-ionization of Fe^{4+} are trapped in ESR-silent defect centres, called X in Fig. 3.9, and thus remain invisible in the wavelength range 650 nm $< \Lambda <$ 850 nm. Only for even smaller wavelengths is the formation of single-hole centres observed.

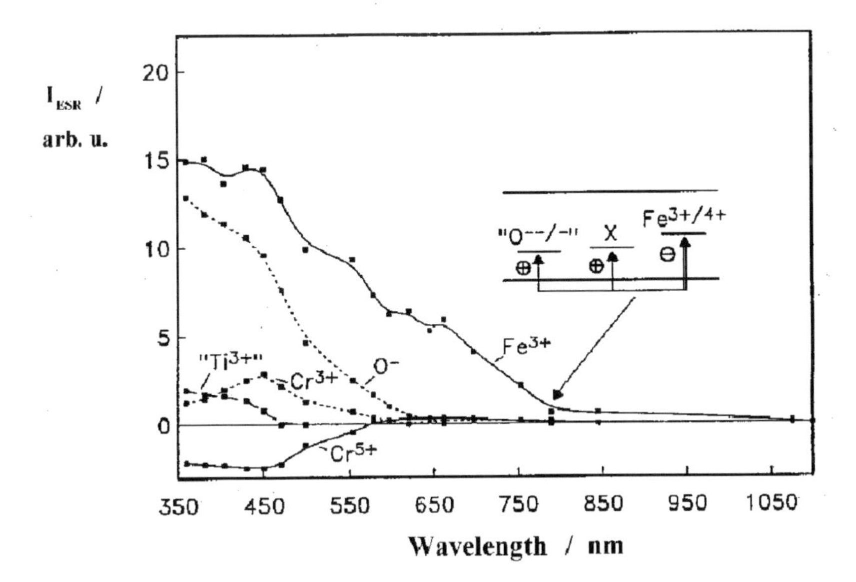

Fig. 3.9. Typical Photo-ESR spectrum obtained from an as-grown $BaTiO_3$ specimen. Changes in the ESR intensity I_{ESR} are shown as a function of the wavelength of the illuminating light. The inset sketches the underlying dominating charge transfer process. Diamagnetic centres X are speculated to correspond to diamagnetic hole bipolarons. The processes do not involve divalent iron, because in as-grown crystals the Fermi level is close to the valence band

A reasonable explanation for the ESR-invisible hole centres has been given by Possenriede et al. [184], which is based on hole bipolarons corresponding to molecular O_2^{2-} aggregates. Similarly to isoelectronic F_2 molecules, such oxygen complexes should possess a diamagnetic ground state, thus being insensitive to ESR. Nearby acceptor defects can be expected to aid the required hole localization. The onset wavelength of single-hole formation in Fig. 3.9 may be interpreted as the light-induced dissociation of bipolarons.

The present subsection is devoted to simulations of trapped hole defects (see also [185, 186, 187]). Shell model and embedded cluster calculations are used to aim at their characterization. In the main, one may think of "on-acceptor" and "off-acceptor" hole trapping. Whereas in the first case the trapped hole localizes at the acceptor site (thereby increasing the acceptor charge state by one positive unit, i.e. $M^{+n}+h^{\bullet} \longrightarrow M^{+(n+1)}$), the hole remains at the oxygen ligands in the second case. Further, the off-acceptor case allows us to distinguish hole states according to their degree of delocalization, i.e. complete localization at exactly one ligand anion, intermediate localization at two oxygen ions (formation of V_k centres), and delocalization over more than two oxygen ligands. V_k defects, which are already known from studying alkali halides [188], can be interpreted as negatively ionized hole bipolarons in the present context.

Which type of localization is favoured depends not only on the ionicity of the host system, but also on the incorporation site and electronic structure of the acceptor point defect. For example, at Ba sites the formation of localized off-acceptor holes can be expected to be more favourable than that of delocalized species due to the pronounced oxygen–oxygen separations. Further, if two holes are bound simultaneously to one acceptor defect the formation of bipolarons will depend on the inter-oxygen binding properties.

Shell-model-based investigations show that single holes are most favourably trapped at Ti site acceptor impurity cations. We first consider the off-acceptor case. It is emphasized that the shell model is accessible only to completely localized holes. Binding energies referring to Ti site acceptors typically range between 0.5 and 1 eV. The binding at Ba site acceptor defects is one order of magnitude less favourable. This prediction agrees with experimental binding energies observed for A_{Ba}^{+}–O_O^{-} complexes (with A being an alkali cation). Monitoring (via ESR) the temperature dependence of the hole emission rate to the valence band yielded energies ≤ 0.1 eV. Similar predictions hold true for trapping of (pre-existing) bipolarons; see Table 3.8.

Tables 3.8 and 3.9 compile the calculated results of single-hole trapping and bipolaron formation at Ti site and Ba site acceptor defects. Single-hole binding energies equal the (negative) second hole-ionization energies reported in Table 3.9. In addition to the shell model parametrization appropriate for $BaTiO_3$ [158] an O^-... O^- pair potential is employed to account for the attractive covalent interaction between two neighbouring O^- hole species assumed to form a hole bipolaron. Corresponding covalency contributions

Table 3.8. Trapping of hole bipolarons at Ti site and Ba site acceptor defects in $BaTiO_3$. Negative binding energies indicate a favourable trapping of bipolarons

Acceptor defect	Bipolaronic bond length / Å	Binding energy (eV) of bipolarons to acceptor type defects
Al^{3+}_{Ti}	+1.24	−1.88
Cr^{3+}_{Ti}	+1.23	−0.61
Mg^{2+}_{Ti}	+1.24	−1.80
Fe^{2+}_{Ti}	+1.22	−0.87
V_{Ti}	+1.19	−1.63
Li^{+}_{Ba}	+1.21	−0.42
Na^{+}_{Ba}	+1.23	+0.03
K^{+}_{Ba}	+2.70	+0.80
V_{Ba}	+2.57	+0.94

Table 3.9. Hole ionization energies of trapped hole bipolarons

Acceptor-type defect	first hole ionization energy (eV)	second hole ionization energy (eV)
Ti^{4+}_{Ti} (isolated bipolaron)	1.25	0
Al^{3+}_{Ti}	2.27	0.85
Cr^{3+}_{Ti}	1.29	0.57
Mg^{2+}_{Ti}	2.19	0.82
Fe^{2+}_{Ti}	1.70	0.42
Li^{+}_{Ba}	1.59	0.08
Na^{+}_{Ba}	1.16	0.06
K^{+}_{Ba}	–	0.12
V^{ll}_{Ba}	–	0.05

are absent in the intrinsic O^{2-}... O^{2-} short-range potential. The O^{-}... O^{-} potential has been generated by simulations of self-trapped holes in corundum based on the INDO approximation [189]. The simulations suggest that trapped stable bipolarons may be interpreted as tightly bound molecules with fixed bond length around 1.2 Å. This calculated bond length is in satisfactory agreement with the bond length ~1.4 Å of isoelectronic free F_2 molecules [190].

Considering trapped hole type bipolarons, three different binding energies can be distinguished: the first describes the binding of bipolarons to acceptor-type defects (corresponding values are compiled in Table 3.8), whereas the second and third measure the affinity of the first and the second hole, respectively, to an acceptor. A complete characterization of the latter two binding energies is given by following ionization reactions:

$$\text{first hole ionization: } A - 2O_O^{-1} \longrightarrow A - O_O^{-1} + O_O^{-1} \quad (3.10)$$

$$\text{second hole ionization: } A - O_O^{-1} \longrightarrow A + O_O^{-1} . \quad (3.11)$$

In these equations A denotes an arbitrary acceptor defect, either on a Ba site or on a Ti site, $A - O_O^{-1}$ represents an acceptor–hole complex and $A - 2O_O^{-1}$, evidently, a hole bipolaron trapped at A. Due to covalency between the O^- ions the first hole ionization energy is in all cases greater than the second hole ionization energy. Therefore, hole-type bipolarons are examples for negative-U defect centres. Table 3.9 summarizes the corresponding hole ionization energies for a number of acceptor defects. It is, finally, emphasized that the stabilization of bipolarons is possible not only due to the additional covalent O^-... O^- interaction, but also because lattice relaxtion aids their formation. This can be shown by simulating two neighbouring O^- holes and omitting any additional covalent interaction. The holes are bound to each other, corresponding to an energy gain of 0.2 eV and to a bond length of 2.69 Å, which is slightly less than the perfect lattice separation of 2.80 Å.

In principle, shell model simulations may also be used to decide between the formation of (localized) off-acceptor and on-acceptor hole states. This is possible upon combining defect formation energies with the appropriate electron affinities of oxygen and the actual cationic ionization potentials. On-acceptor holes are certainly favourable, if the acceptor cation possesses filled electron levels above the valence band edge. Rough estimates are obtained with free-ion ionization potentials. Whereas in the case of monovalent alkali cations (Ba sites), divalent magnesium and trivalent aluminium (both at Ti sites) the formation of off-acceptor holes is significantly preferred, on-acceptor holes probably occur at many divalent transition metal cations like Fe^{2+} and Mn^{2+} which are incorporated at Ti sites. The latter predictions are even further stabilized upon inclusion of crystal field splitting energies. However, the results remain particularly uncertain for higher-valent transition metal impurities. For example, the calculations predict Fe_{Ti}^{3+}–O_O^- to be more favourable than Fe_{Ti}^{4+}, corresponding to an energy gain of 4 eV. Referring to the discussion in Sect. 3.2 this result may be easily reversed due to covalent charge transfer and crystal field effects which, acting constructively in this situation, could shift the e_g levels of iron above the valence band states. It is noted, in addition, that there are no experimental indications in favour of the dipolar iron–hole complexes.

Despite the numerical uncertainties, the shell model-based considerations suggest that increasing the ionicity of the defect complex probably favours the formation of off-acceptor holes.

Further insight into questions referring to trapped holes is gained by embedded cluster calculations. Corresponding investigations are indispensable in order to aim at unambiguous results regarding the precise localization properties of single trapped holes and the formation of possible (diamagnetic) hole bipolarons. Subsequently, particular attention will also be paid to the hole-localization properties at higher-valent transition metal acceptor cations.

In all subsequently discussed embedded cluster studies the central quantum defect cluster is chosen analogously to the simulations of manganese reported in the preceding subsection (see Fig. 3.3). Thus, we confine our considerations to Ti site acceptor defects. The MO ansatz for the central acceptor–oxygen complex MO_6 employs Gaussian-type basis functions with split valence (SV) quality for the acceptor impurity (Mg^{2+}, Al^{3+}, Cr^{3+}, Fe^{3+} and Rh^{3+}) and its oxygen ligands; the basis set is further augmented by polarizing d-functions. Bare effective core potentials are used to model ion-size effects of the outer Ba and Ti cations.

The calculations employ HF theory, and electron correlation is included on the basis of SDCI(+Q), MP2 and density functional theory. In DFT calculations two choices are used in order to approximate the unknown exact exchange correlation functional, i.e. the local spin density ansatz of Vosko, Wilk and Nusair [191] (VWN-LSDA) and the advanced "BLYP" functional (see Sect. 2.1), which improves on L(S)DA.

In order to perform cluster geometry optimizations which are consistent with the embedding crystal lattice (represented by a point charge field) an additional program has been written which updates the total cluster energies and gradients, as calculated by any quantum chemical program such as CADPAC [173], by adding appropriate short-range pair potential contributions due to the required interactions between cluster ions and embedding lattice ions. With these updates the program carries through the cluster geometry optimization using a variable metric (quasi-Newton) minimization algorithm. Relaxations of the embedding shell model lattice are determined using the CASCADE computer program [109]. Details of the complete procedure have been outlined in Sect. 2.1.2.

During the lattice equilibration step (Sect. 2.1.2) the ion charges referring to the MO_6 acceptor–oxygen complex are chosen according to the intended hole localization, e.g. completely localized off-acceptor holes are given as O_O^{-1}, bipolarons, correspondingly, as a pair of localized off-acceptor holes, and the V_k centre as a pair of $O_O^{-1/2}$ species. All remaining ions which are not affected by holes retain their formal integral charges. This simplifying choice of point charge representations can be justified by Mulliken population analyses (MPA). Table 3.10, for example, presents the MPA ionic charges obtained for the equilibrated bipolaron singlet state (HF theory). Moreover, the dipole and quadrupole moments of the cluster calculated on the basis of formal ionic point charges are in reasonable agreement with the respective moments

Table 3.10. MPA-derived ion charges referring to the equilibrium bipolaron singlet HF state. O^{2-}_{plane} denotes the oxygen ligands within the plane of the hole bipolaron O^{-}–O^{-} and $O^{2-}_{perp.}$ correspondingly denotes the oxygen ions along the axis perpendicular to the bipolaron plane

Ion	Ionic MPA charge
Mg^{2+}	+1.46
O^{2-}_{plane}	−1.88
$O^{2-}_{perp.}$	−1.89
O^{-}	−0.95

of the quantum cluster. Whereas the dipole moments of the point charge cluster deviate by about 2% from their quantum analogues, the quadrupole moments are reproduced in the point charge approximation with differences being slightly less than 6%. Therefore the electrostatic potential generated by the cluster charge density will be sufficiently reproduced when employing the point charge substitute.

First, we discuss embedded cluster calculations for trapped single holes. Hole states corresponding to different localization properties are considered. For all investigated Ti site acceptors HF theory predicts localized off-acceptor holes to be highly preferred. This result complies with predictions derived from the shell model simulations discussed above. Table 3.11 compiles the HF and MP2 calculations for trivalent transition metal cations. The formation of localized (i.e. symmetry-broken) off-acceptor holes in π-type oxygen 2p-orbitals is preferred. It is noted that the symmetry breaking is stabilized by defect-induced lattice distortions. Therefore these solutions emerging from the so-called HF symmetry dilemma [182, 183] may attain physical significance. For holes occupying π-type oxygen orbitals the spin coupling to the iron cation is negligible. σ-type holes, on the other hand, are about 1 eV less favourable. Figure 3.10 visualizes both localized off-acceptor hole states.

In the case of σ-type localized holes we infer from Table 3.11 a small spin-coupling interaction (0.1 eV) favouring antiparallel spin alignment between iron and O^{-}. The ordering is explained upon observing that for $2S+1=7$ the bonding σ-orbitals tend to increase the local spin density at the iron site which, however, is unfavourable due to the Pauli principle (note that iron has a half-filled 3d-subshell corresponding to high spin).

Delocalized off-acceptor holes as well as tetravalent iron are highly unfavourable within HF theory. The delocalization properties of both these states are very similar, and thus energetic differences turn out to be negligible. Analogous results are obtained for different transition metal cations like Cr^{3+} and Rh^{3+}; in particular, the symmetry-broken solutions Cr^{3+}_{Ti}–O^{-}_{O} and Rh^{3+}_{Ti}–O^{-}_{O} are preferred over the actual on-acceptor states. Moreover, delocalized off-acceptor hole states are unfavourable.

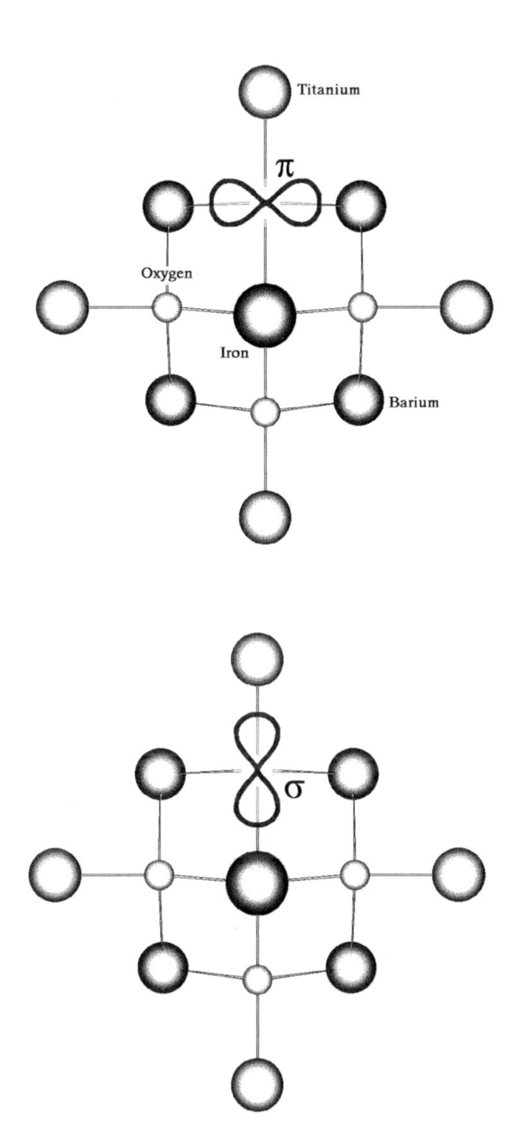

Fig. 3.10. Visualization of localized off-acceptor π- and σ-holes close to an $\mathrm{Fe}_{\mathrm{Ti}}^{3+}$ impurity. The symmetry-broken states are stabilized by lattice relaxations

Table 3.11. Total HF and MP2 embedded cluster energies for different hole states including lattice relaxations. Energies are renormalized to localized off-acceptor states. $O_O^-(\pi, \sigma)$: localized off-acceptor hole in oxygen 2p π- or σ-type orbitals. $(O(xy))^-$: off-acceptor hole delocalized over the four oxygen ligands in the xy-plane. $2S+1$ denotes the total spin multiplicity of the cluster. Note that Fe^{3+} and Cr^{3+} favour high spin ($2S_{Fe}+1=6$, $2S_{Cr}+1=4$), and Rh^{4+} prefers low spin ($2S_{Rh}+1=2$)

Defect	$2\hat{S}+1$	ΔE_{total}^{UHF}(eV)	ΔE_{total}^{MP2}(eV)
$Fe_{Ti}^{3+} - O_O^-(\pi)$	7	+0.0	+0.0
$Fe_{Ti}^{3+} - O_O^-(\pi)$	5	+0.0	–
$Fe_{Ti}^{3+} - O_O^-(\sigma)$	7	+1.1	–
$Fe_{Ti}^{3+} - O_O^-(\sigma)$	5	+1.0	–
$Fe_{Ti}^{3+} - (O(xy))^-$	7	+2.4	+0.5
Fe_{Ti}^{4+}	5	+2.4	+0.1
$Rh_{Ti}^{3+} - O_O^-(\pi)$	2	+0.0	+0.0
$Rh_{Ti}^{3+} - (O(xy))^-$	2	+2.7	+0.1
Rh_{Ti}^{4+}	2	+0.4	−2.4
$Cr_{Ti}^{3+} - O_O^-(\pi)$	5	+0.0	+0.0
Cr_{Ti}^{4+}	2	+1.15	−2.6

In comparison to HF theory the inclusion of electron correlation increases the covalent charge transfer which is directed from the oxygen ligands onto the central metal cation. As a result the ligand anions become less negatively charged, whereas the positive cation charge is reduced at the same time. Thus, electron correlations stabilize the (remaining) electrons at anions, but destabilize those located at the acceptor cation. There are two important consequences of the additional correlation-induced charge transfer: first, the hole localization at the acceptor cation becomes enhanced, and, second, off-acceptor holes increasingly delocalize, because ionization of exactly one ligand anion becomes unfavourable due to the correlation-induced enhancement of the oxygen electron affinity. In summary, HF theory models $BaTiO_3$ as too ionic, and this artificially favours the formation of symmetry-broken localized off-acceptor hole states in all cases studied so far. This defect becomes removed upon inclusion of electron correlations. Possible situations where HF theory could provide reliable predictions might refer to divalent transition metal acceptors at Ti sites. Here, one may expect that on-acceptor holes are most favourable. Divalent copper impurities are of particular interest: preliminary embedded cluster calculations [187] confirm the formation of on-acceptor holes in this case. Obviously, the situation differs from high-T_c oxides, where doped holes are predicted to enter the oxygen sites. Major differences could emerge from the fact that divalent copper introduces a doubly negatively charged acceptor defect in $BaTiO_3$, but not in the copper-based superconducting oxides, where it behaves neutrally.

MP2 calculations (see Table 3.11) confirm that the inclusion of electron correlation tends to favour the formation of on-acceptor hole states. Further, considering off-acceptor holes, correlation mostly affects the delocalized hole states, but affects localized states to a much lesser extent. These results, referring to off-acceptor holes, demonstrate the pronounced interplay between orbital relaxations and electron correlation, i.e. the localized hole states with significant HF orbital relaxations receive less correlation energy gain than delocalized hole states. Whereas Fe^{4+} remains 0.1 eV less favourable than $Fe^{3+} - O^-$, Rh^{4+} and Cr^{4+} become clearly preferred over the particular $M - O^-$ centres by 2.4 eV and 2.6 eV, respectively. Finally, employing DFT-BLYP proves even Fe^{4+} to represent the most favourable hole-trapping state. Moreover, for all simulated transition metal acceptors delocalized hole states represent the most favourable off-acceptor holes at this correlated level at least. However, the energy separations between localized and delocalized off-acceptor holes remain within some tenths eV. For iron this result indicates that the localized off-acceptor hole would be slightly more favourable than its delocalized counterpart if the ionicity of Fe^{3+} could be increased. In practice, this condition might be satisfied for suitable platinum cations which possess a partially filled 5d-subshell. Due to the expected lanthanide contraction the platinum impurities could behave ionically, leading to observable localized off-acceptor holes, if these are sufficiently stable or metastable against the formation of on-acceptor holes. Indeed, based on ESR such a localized off-acceptor hole defect bound to a Pt-impurity has been proposed to exist [192]. The corresponding information was obtained by analyzing the hyperfine interactions of the ESR-active hole state. The platinum cations, on the other hand, are ESR silent. The charge state of platinum could not be assessed experimentally, but Pt^{4+} having a completely filled $5d(t_{2g})$ subshell represents a possible candidate.

In summary, it is essentially electron correlation and not lattice relaxation that stabilizes the high-valent charge states of various transition metal cations in $BaTiO_3$. This mechanism resembles the impurity stabilization in semiconducting materials corresponding to the model of Haldane and Anderson [7]. On the other hand, in highly ionic materials we should expect the stabilization of impurity charge states mainly due to lattice relaxations [193]. The present discussion confirms that $BaTiO_3$ represents a semi-ionic material.

Up to this stage we have discussed the formation properties of trapped holes near to transition metal cations. Covalency leads to favouring the on-acceptor hole states. But there is also a number of acceptor type impurity cations behaving ionically. Examples are the monovalent alkali cations, divalent magnesium and trivalent aluminium. Obviously, on-acceptor hole states will be unfavourable. Since covalent charge transfer occurs to a much lesser extent, one might expect the formation of localized off-acceptor holes to be favourable. The subsequent calculations, compiled in Table 3.12, exemplify the situation for Al^{3+}_{Ti} acceptor cations. ESR investigations have shown that the formation of V_k hole centres is favourable in this case [194], but note that there are no corresponding indications for other acceptor cations.

Table 3.12. Hole formations at Al^{3+}_{Ti}. All total embedded cluster energies are renormalized with respect to the localized off-acceptor π-type hole state, which represents the HF ground state. Within DFT-BLYP this state can only be stabilized when using the relaxed HF cluster configuration. DFT energies refer to this intermediate state (denoted by an asterisk). However, the localized hole is found to be unstable upon further geometry optimization, during which it changes into a delocalized hole state

Defect	ΔE^{UHF}_{total}(eV)	ΔE^{MP2}_{total}(eV)	$\Delta E^{DFT-BLYP}_{total}$(eV)
$Al^{3+}_{Ti} - O^{-}_{O}(\pi)$	+0.0	+0.0	+0.0*
Al^{3+}_{Ti} – (deloc.(π))$^-$	+2.5	+0.4	−0.8
Al^{3+}_{Ti} – (deloc.(σ))$^-$	+3.7	+1.9	–
Al^{3+}_{Ti} – ($V_k(\pi)$)	+1.0	+0.0	−0.4
Al^{3+}_{Ti} – ($V_k(\sigma)$)	+2.8	+1.6	–

Within HF theory the ground state is given by the localized off-acceptor π-type hole state. The V_k π-type hole distribution (Fig. 3.11) is less favourable by 1 eV.

Figure 3.12 displays the calculated hole spin densities for both V_k modifications.

The introduction of electron correlation again increases the delocalization of holes. The correlated calculations reported in Table 3.12 confirm that all hole states are rather similar in energy. Most favourable are the delocalized state and the formation of V_k centres, but unfortunately there is no clear indication in favour of the latter hole states. The calculated results suggest the existence of a delicate balance between the local electronic structure and defect-induced lattice distortions: increasing the acceptor ionicity would result in the formation of completely localized off-acceptor holes, whereas a reduced ionicity leads to delocalization of hole states. In an intermediate small "window" the formation of V_k centres should be most favourable. The ESR data confirm that aluminium acceptors fall into this window, but the present embedded cluster model also proves to be close to this intermediate situation. Further influences affecting the possible formation of V_k centres seem to be related to the local electronic structure of the acceptor cation. For example, the present embedded cluster calculations give no hints about the

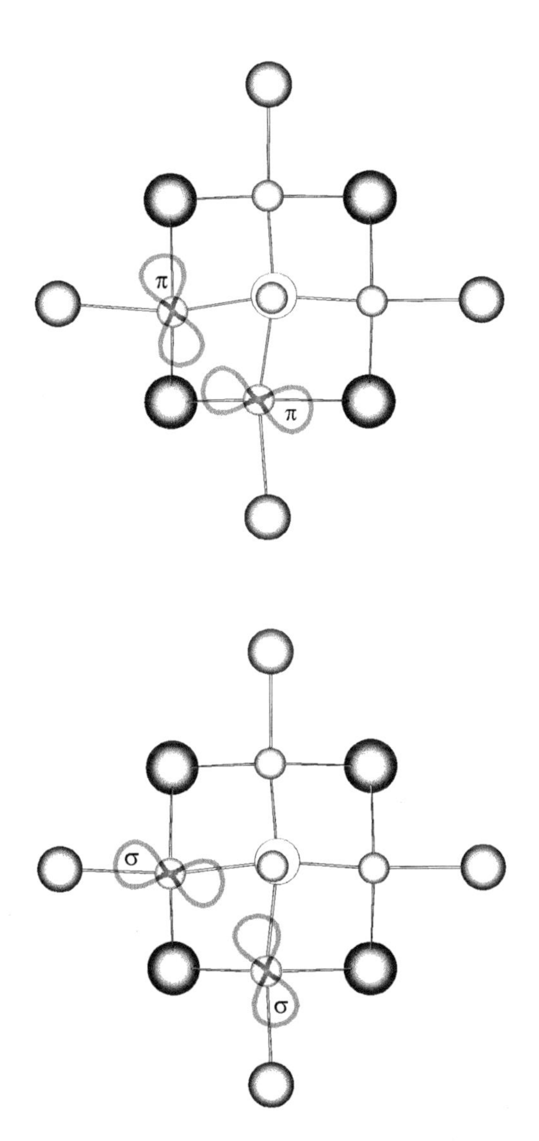

Fig. 3.11. Visualization of π- and σ-type V_k hole centres trapped by an Al^{3+}_{Ti} acceptor cation. The interatomic separation between the oxygen partners forming the V_k defect is 2.3 Å and 2.6 Å for hole localization in π- and σ-type oxygen orbitals, respectively. In comparison the perfect-lattice separation between neighbouring oxygen ions is 2.8 Å

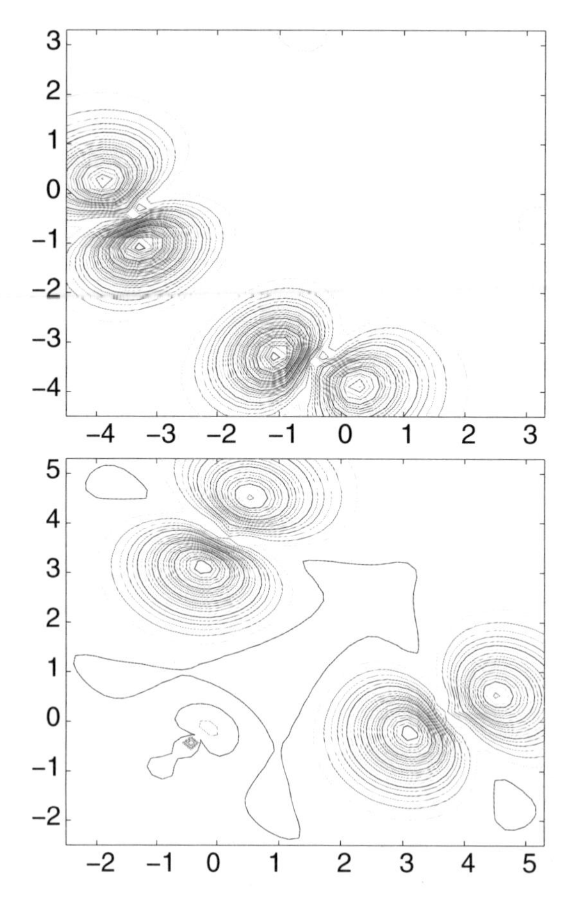

Fig. 3.12. Hole spin density plot of π- and σ-type V_k centres trapped at Al^{3+}_{Ti}. The occupation of oxygen $2p$ orbitals can be clearly seen. The acceptor is located at $(x, y) = (0, 0)$ in the upper picture (π-type), and at (−0.5,−0.5) in the lower picture (σ-type). All xy-coordinates in the figures are given in atomic units. The interatomic separations between the oxygen partners forming the V_k defect are 2.3 Å and 2.6 Å for localization in the π- and σ-type oxygen orbitals, respectively. The perfect lattice separation is 2.8 Å

formation of V_k defects for transition metal acceptors which are characterized by incompletely filled 3d-subshells.

Returning to aluminium, the σ-type modification of the V_k centres is by 1.6–1.8 eV less favourable than the π-type V_k defect. Moreover, all ion displacements are smaller in the σ-type situation. This result receives particular importance remembering that V_k hole centres may be interpreted as negatively ionized hole bipolarons. Therefore we may expect the same relaxation behaviour considering π-type and σ-type hole bipolarons. Due to the reduced lattice relaxations, the latter species would be more appropriate to coherent motions. It is noticed that holes created upon doping within the Cu–

O planes of high-T_c superconducting oxides are assumed to populate σ-type oxygen orbitals [195, 196]. Therefore we should expect the same behaviour for bipolarons – if they exist in these materials. σ-type bipolarons might be speculated to facilitate the observed high-temperature superconductivity, thereby supporting the scenario proposed by Alexandrov and Mott [8].

It is instructive to compare the present embedded cluster calculations with earlier investigations of single holes trapped at Li^+-acceptor cations in MgO [197, 198] and NiO [199]. All of these calculations have been based on HF theory, and thus predict symmetry-broken localized off-acceptor holes to be most favourable. However, in reference [198] the effect of electron correlations has been tested on the basis of preliminary SDCI. It was found that correlation reduces the hole diffusion energy barrier from 1.05 eV (HF) to 0.23 eV. Therefore the picture emerging from these investigations seems to comply with the correlation-induced delocalization of hole states found in the present embedded cluster calculations. The possible formation of hole-type V_k centres has been investigated in corundum [200]. Also, these simulations were confined to the HF level of approximation. Based on the present experience these calculations would benefit from additional simulations at the correlated level.

We now turn to simulations of trapped hole bipolarons in $BaTiO_3$. As has been stated already, these defect species are closely related to the V_k hole centres. Divalent magnesium substituting for titanium is chosen as the actual acceptor cation. Correspondingly, the magnesium–bipolaron complex defines a neutral lattice perturbation. Two geometrical hole–acceptor configurations will be considered: the bipolaron with two holes on neighbouring oxygen ions trapped at the magnesium impurity and a linear complex $O^- - Mg^{2+}_{Ti} - O^-$. The presence of two holes introduces a natural driving force towards complete localization, of which the linear complex corresponds to minimizing the inter-hole Coulomb repulsion. The bipolaron results from lattice relaxation and covalent bonding interactions. Both hole complexes are studied under different conditions:

- Hartree–Fock (HF) treatment of the quantum defect cluster employing a rigid crystal lattice with ions held fixed on their perfect lattice positions. Only the actual O^- partners are allowed to relax.
- HF description of the cluster including complete lattice relaxation.
- SDCI+Q and MP2 investigations of the central defect cluster using the equilibrium lattice of the previous step. The computing costs impede geometry optimization in practice. Further geometry optimizations are performed only at the DFT level (LSDA and BLYP).

Figure 3.13 displays the perfect lattice results. First, we observe that the triplet bipolaron state is energetically more favourable than the singlet state. The reason for this behaviour may be traced back to electronic interactions between the O^- ions and crystal ions in the immediate neighbourhood. In particular, the triplet bipolaron takes advantage of this interaction because of its antisymmetrical molecular charge distribution. Increasing the bipolaron bond length destabilizes the bipolaron due to increasing delocalization of the holes over all oxygen ligands. This effect is more pronounced for the triplet state. Thus, at this level, the hole bipolaron can be considered as an embedded molecular dimer only close to its singlet equilibrium separation. The energetically most favourable configuration, however, corresponds to the linear complex with localization of the holes at two adjacent oxygen anions. This result should be expected from our previous discussion of single holes, as HF simulations place a disadvantage on the formation of delocalized states. For the purpose of comparison, results referring to the linear complex are included in Fig. 3.13. The linear configuration is ~1 eV more favourable than the triplet bipolaron state. It is finally noted that there is no significant energy spacing between the singlet and triplet states of the linear complex because of negligible spin coupling in this case.

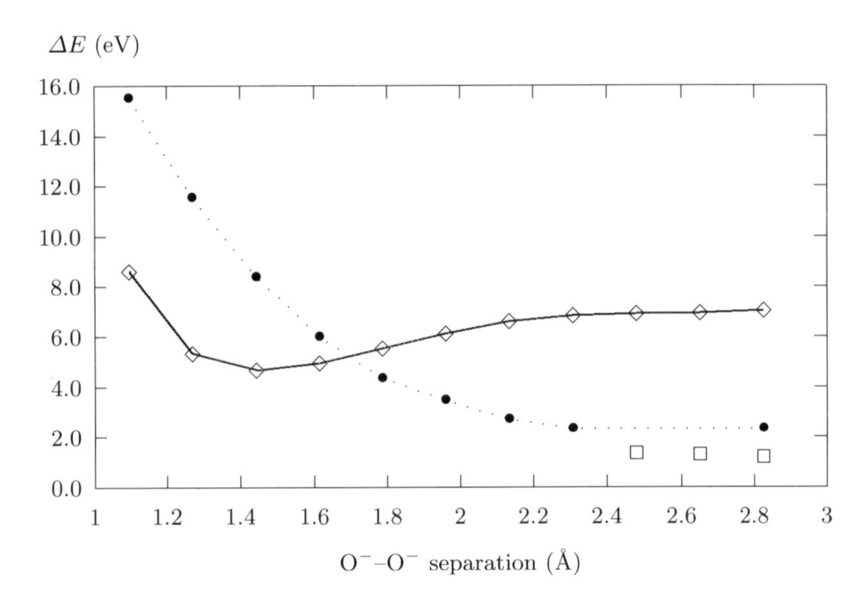

Fig. 3.13. Energy dependence of the bipolaron singlet (◇) and triplet (•) state as a function of the O^-–O^- bond length employing a perfect (thus unrelaxed) crystal lattice and the HF approximation. The additional boxes (□) correspond to the triplet state of the linear defect complex, into which the HF calculations converged in these cases, although they started with converged (triplet) bipolaron orbitals corresponding to a previous bipolaron configuration

Figure 3.14 displays the HF energy dependences of the singlet and triplet bipolaron states, including the effects of lattice relaxation, for which the general pattern is similar to that shown in Fig. 3.11 for the V_k centre. However, due to the stronger bonding effects, the O^- displacements are considerably more pronounced. It is noted that bonding is accomplished by π-type oxygen 2p orbitals. A computationally effective but simplified procedure based on pair potential derived cluster relaxations has been employed to generate Fig. 3.14. But most importantly, this figure reflects the relevant extremum features of exact HF-based geometry optimizations: the equilibrium bipolaron bond length is found to be 1.44 Å, which is also in good agreement with the shell model simulations discussed above. Further, at its equilibrium separation the diamagnetic singlet state represents the most favourable bipolaron state. Increasing the bond length still drops the triplet state below the singlet state, but in contrast with the perfect lattice situation the equilibrium singlet state remains more favourable by 1.71 eV than the minimum triplet state. The triplet state assumes its equilibrium separation at 1.96 Å.

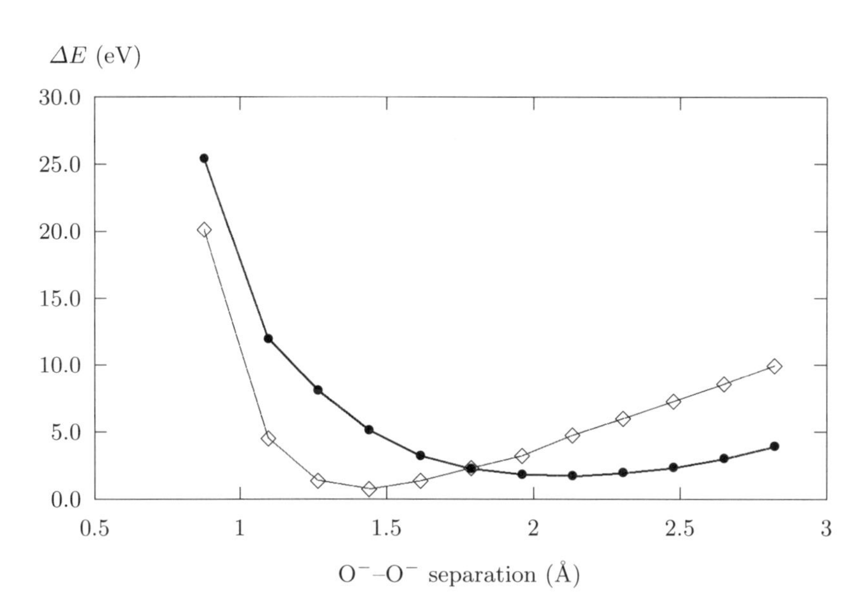

Fig. 3.14. Energy dependence of the bipolaron singlet (◇) and triplet (•) state as a function of the O^-–O^- bond length employing a relaxed crystal lattice and the HF approximation. Though lattice relaxation has been treated approximately, the figure reproduces the important features of exact geometry optimizations (see text)

However, within HF the linear configuration is more favourable by ~1.9 eV than the competitve hole bipolaron. Thus, the simulated bipolaron represents only a metastable defect complex at this quantum mechanical level. Despite the established disfavour of hole bipolarons, as found within HF, we can infer a general trend of defect-induced crystal distortions towards stabilization of singlet-state bipolarons. This may readily be seen by comparing different stages of a relaxation: within a perfect lattice the energy difference between the bipolaron (with the bipolaron-forming oxygen ions at neighbouring perfect lattice positions) and the linear complex is ~6 eV. This is reduced to about 3.5 eV upon relaxing the O^- ions with all other cluster and embedding lattice ions remaining fixed at their perfect lattice positions. Finally, the inclusion of complete lattice relaxation yields 1.9 eV. The stabilizing effect of lattice relaxations relates to a localization of the holes: as has been stated above, the bipolaron hole states become increasingly delocalized upon separating the corresponding oxygen partners within an otherwise perfectly structured crystal. On the other hand, Mulliken population analyses show that the hole states remain significantly more localized at each O^-–O^- separation if complete relaxation of the embedding lattice is taken into account. The hole bipolaron receives more similarity to an embedded dimer. Consequently, defect-induced crystal relaxations also tend to favour the spin-singlet state of bipolarons over the spin-triplet state.

All HF-based simulations consistently show that hole-type bipolarons are unstable (or metastable) within HF theory. However, this particular result should not surprise, because HF is known to underestimate the molecular bonding properties. It is instructive to compare the present bipolaron simulations with calculations on related isolated dimers, i.e. for the isoelectronic F_2 and O_2^{2-} molecules. The predicted instability of isolated F_2 within HF theory (see also [201], for example) parallels the findings for bipolarons. The HF binding energy of the fluorine molecule is +1.48 eV. HF potential energy curves (PEC) for F_2 and (isolated) O_2^{2-} are shown in Figs. 3.15 and 3.16, respectively.

Due to the unscreened Coulomb repulsion between the oxygen ions the energy decreases with increasing bond length. It is recalled in this context that it may become dangerous to imply restrictions due to the molecular symmetry in the determination of one-electron orbitals, which are used to construct the all-electron HF wavefunction. Such symmetry-conserving calculations may well provide the wrong dissociation states, which is a manifestation of the so-called symmetry dilemma of HF theory [182, 183]. Corresponding effects are exemplified by the "□"-marked curve in Fig. 3.16. Artificially, this curve yields a binding energy of −2.9 eV for O_2^{2-}. Moreover, there is no Coulomb repulsion between the two O^- anions using this symmetry-restricted solution. The non-restricted "◇" curve in Fig. 3.16 yields the correct dissociation behaviour; as expected this spin-singlet state agrees with the spin-triplet state (+) at large separations. The HF-based dissociation energies of various molecular oxygen anions quoted in reference [202] are related to corresponding symmetry-restricted calculations involving the wrong dissociation state, which is built up by symmetry-restricted delocalized one-electron orbitals. For example, the dissociation energy of O_2^- is not 10.6 eV, based on restricted HF, but 1.0 eV, as derived from the symmetry-broken one-determinant solution to the ground state. Similarly, the symmetry-restricted HF calculations for F_2 simulate a dissociation energy of +8.9 eV instead

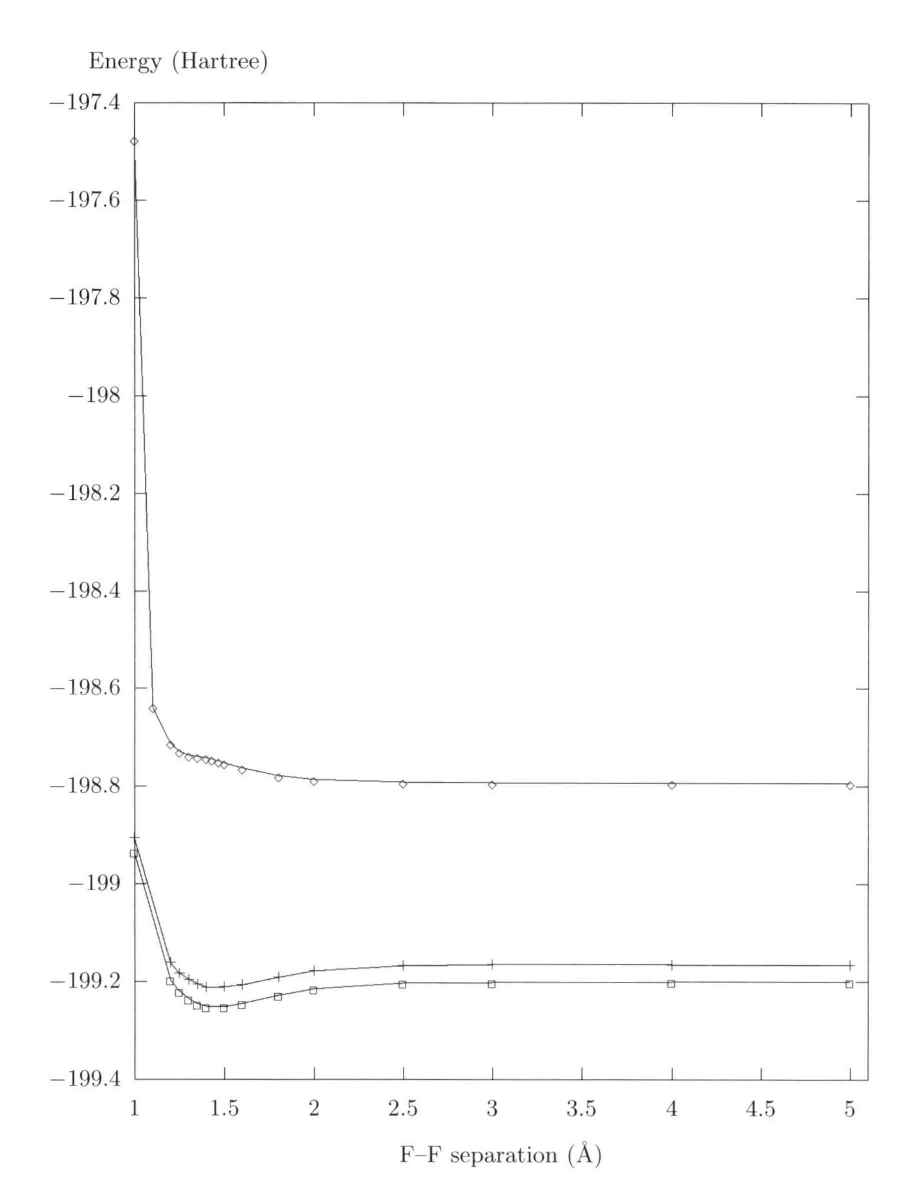

Fig. 3.15. Potential energy curves (spin-singlet ground state) for isolated F_2 molecules: symmetry-broken Hartree–Fock (◇), MRSDCI (+) and MRSDCI+Q (□)

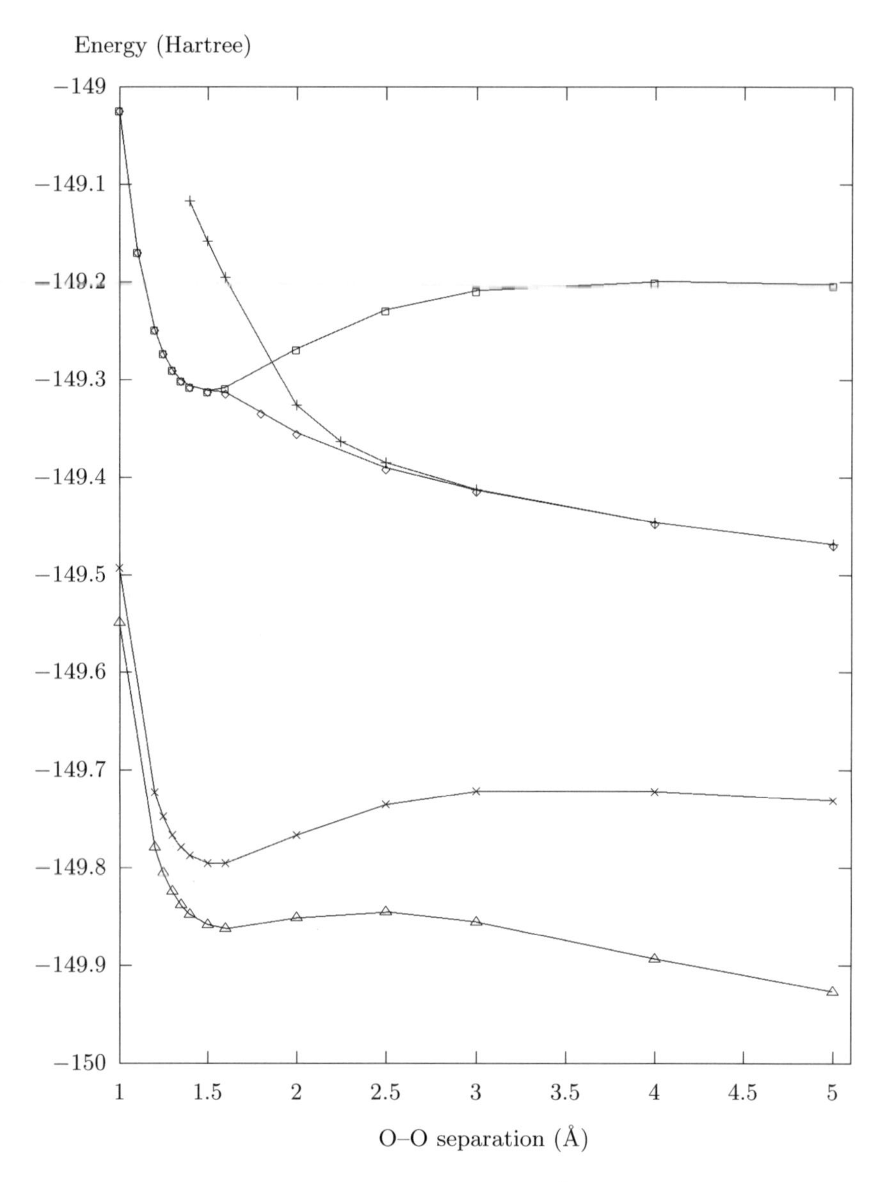

Fig. 3.16. Potential energy curves for isolated O_2^{2-} molecular anions. Symmetry-broken spin-singlet and spin-triplet curves using Hartree–Fock (◇ and +, respectively), singlet state Hartree–Fock curve based on a symmetry-restricted one-determinant wavefunction (□), and singlet state curves including correlations: MRSDCI (×) and MRSDCI+Q (△)

of the correct -1.48 eV. As a matter of fact, the exact molecular wavefunctions must possess the symmetry of the considered molecule. Therefore, wavefunction-based *ab initio* descriptions should provide schemes in order to obtain symmetrized wavefunctions. In the present situation this could be accomplished by the proper superposition of symmetry-broken one-determinant wavefunctions. This method goes beyond HF theory and represents a simple example of the non-orthogonal configuration interaction approach [203]. However, within HF theory the symmetry-broken solutions are sufficient to obtain reliable estimates of molecular potential energy curves.

The similarity of F_2 and *embedded* O_2^{2-} molecules suggests that the relaxed lattice effectively screens the O^- charges. Anticipating this similarity between F_2 and *embedded* O_2^{2-}, it appears not very surprising that HF theory is unable to predict a stable bipolaron state. It is therefore mandatory to include electron correlations.

We first consider the isolated molecular species. PECs for isolated F_2 and O_2^{2-}, which were obtained from multi-reference configuration interaction calculations employing single and double electronic excitations (MRSDCI+Q), are also displayed in Figs. 3.15 and 3.16. The "+Q" denotes corrections to the "bare" MRSDCI results in order to guarantee size consistency. This is achieved on the basis of the Davidson formula (see Sect. 3.3.1). Whereas for F_2 the '+Q' correction is of minor importance, it becomes necessary in the calculations for O_2^{2-}. Generally, the neglect of size consistency corrections leads to an underestimation of correlations at larger separations and, correspondingly, to an overestimation of binding energies (compare the "$\times$" and the "$\triangle$" curves in Fig. 3.16, for instance).

In the case of the fluorine molecule the calculated bond length corresponds to 1.45 Å and the binding energy is -1.4 eV (the corresponding experimental values are 1.41 Å and -1.7 eV). In comparison, density functional theory (DFT) produces the following numbers:

- VWN-LSDA: r_B=1.39 Å, E_B=-3.4 eV ;
- BLYP: r_B=1.43 Å, E_B=-2.2 eV .

These results show that LSDA in particular overestimates molecular binding energies. The results are significantly improved by applying the GGA-type BLYP exchange correlation functional.

For isolated peroxy anions the CI calculations predict a local minimum corresponding to a metastable state close to 1.5 Å. At larger oxygen–oxygen separations the Coulomb repulsion overcomes the bonding effects. The MRSDCI+Q curve is parallel to the correct HF curve in this regime.

It is now instructive to discuss the corresponding results for the embedded O_2^{2-} peroxy molecules. The binding energy of hole bipolarons can be estimated by the energy difference $\Delta := E(\mathrm{BP}) - E(\mathrm{O}^- - \mathrm{Mg} - \mathrm{O}^-)$, since, based on shell model results, the linear $\mathrm{O}^- - \mathrm{Mg} - \mathrm{O}^-$ complex is only slightly bound in $BaTiO_3$. The qualitative trends which have already been found for the isolated molecular species remain true in the case of the bipolarons:

- SDCI+Q and MP2: $\Delta = -0.41$ eV ;
- VWN-LSDA: r_B=1.48 Å, $\Delta = -2.1$ eV ;
- BLYP: r_B=1.55 Å, $\Delta = -1.13$ eV .

With the exception of the SDCI+Q (which are confined to single-reference calculations in the case of the MgO_6 complex) and MP2 simulations, which are based on the relaxed HF cluster geometry, all DFT calculations employ their consistently relaxed crystal lattices. The above results provide a similar ordering of the correlation treatments, as was found for the isolated molecules. The quantum chemical descriptions of correlations (SDCI+Q and MP2) approach the exact binding energy from below, whereas BLYP and, particularly, LSDA are expected to overestimate the bonding correlations. The correct binding energy is supposed to be closer to BLYP than to MP2, because MP2 as well as the SDCI+Q results may well underestimate electron correlations within the MgO_6 complex. Both MP2 and SDCI+Q predict a correlation energy gain of about 2 eV in favour of hole-type bipolarons.

In conclusion, the present investigations demonstrate that defect-induced crystal relaxation and electronic correlations are necessary to stabilize trapped hole bipolarons in $BaTiO_3$. Crystal relaxations increase the localization of bipolaronic hole states and lead to a spin-singlet ground state. The relaxations of the host lattice enable the idea of an embedded O_2^{2-} molecule with electronic properties similar to the isolated F_2 dimer. In particular, HF theory predicts both species to be unstable. In both cases the ultimately found stability is determined by the electronic correlation contributions to molecular bonding.

Although we have only discussed the formation of trapped hole bipolarons in $BaTiO_3$, one may speculate that paired hole species are of general importance in any oxide material. Possible differences will probably refer to the occupation of π- or σ-type oxygen 2p orbitals. In passing we remark again that the energy separation between singlet and triplet bipolarons depends on the induced lattice relaxations. Whereas in perfect lattices the triplet state becomes preferred, the singlet state is most favourable upon inclusion of complete lattice distortions. In this latter situation the “spin gap” is found to be 1.50 eV using MP2 (1.71 eV within HF) for π-type bipolarons. In their bipolaron-based theory of high-temperature superconductivity, Mott and Alexandrov suggested a spin gap of a few tens meV [8]. Such a small value might indicate that lattice relaxations would be present in high-T_c oxides, but to lesser extent than in $BaTiO_3$; this observation is important if coherent motion of bipolarons is to be required. Future work should be devoted to the question of whether σ-type bipolarons (compare the discussion of V_k centres) fit in with this qualitative expectation and with the proposed spin gap.

Analogously, one may expect that electron-type bipolarons are of comparable significance if conduction band states become populated. For example, electron bipolarons have been found to exist in $LiNbO_3$ (see Chap. 5). The following subsection is devoted to preliminary embedded cluster calculations on electron bipolarons in $BaTiO_3$.

3.3.3 Simulation of Electron-Type Bipolarons

The existence of electron bipolarons in $BaTiO_3$ has been proposed upon observing that the paramagnetic susceptibility (due to Ti_{Ti}^{3+} small polarons) of reduced $BaTiO_3$ is smaller by one or two orders of magnitude than expected [204]. Consequently, a significant proportion of small polarons is trapped to form diamagnetic defect centres, of which bipolarons would constitute one natural choice.

In this subsection we shall briefly discuss the results of some preliminary embedded cluster calculations, which aim to explore the possibility of such defects in (reduced) $BaTiO_3$. The investigated cluster is visualized in Fig. 3.17. It corresponds to modelling a bipolaron which is orientated along the [110] direction.

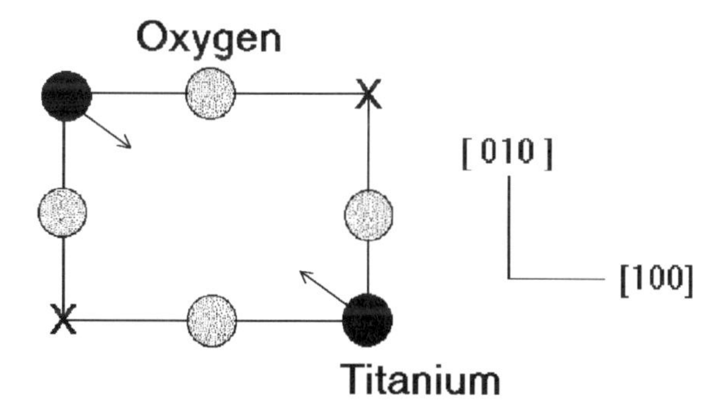

Fig. 3.17. Visualization of the quantum cluster which is employed to simulate electron bipolarons. Black and grey shaded circles denote Ti cations and oxygen anions. These ions are treated quantum mechanically. The "X" represent point charge Ti cations, and a point charge representation is also used to model the embedding crystal lattice. Arrows denote the direction of titanium dispacements upon bipolaron formation

The employed Gaussian-type basis set is chosen to include 4s and 4p orbitals for titanium [125]. Moreover, the split valence basis set of oxygen [126] is augmented by additional polarizing d-functions. Cluster–lattice interactions are mediated by the known pair potentials for $BaTiO_3$. The quantum mechanical level of these embedded cluster calculations corresponds to DFT-BLYP using the CADPAC code [173].

In order to check the compatibility between the cluster and its embedding lattice, a geometry optimization is performed for a defect-free cluster without any polaron-forming electrons occupying titanium 3d-states. The resulting ion displacements due to cluster–lattice mismatch are less than 0.05 Å. Having established the usefulness of this preliminary cluster representation, a bipolaron is studied upon adding two additional electrons which localize at the two quantum mechanically simulated Ti cations. Assuming a spin-singlet state, both Ti cations approach each other by almost 0.3 Å (compare the arrows in Fig. 3.17), thus yielding a bond length of 2.2 Å. Additional simulations probe the bonding contribution of the 4s- and 4p-type titanium orbitals. Calculations in which only the titanium cations are allowed to relax with all other crystal ions held fixed at their perfect lattice positions, indicate that the bonding between the two Ti-polarons is facilitated by significant admixtures of these states to the occupied $3d_{xy}$ orbitals[4] leading to a hydrogen-like bonding: after subtracting the 4s and 4p orbitals the separation between both titanium cations increases by about 0.2 Å, and, with respect to the perfect-lattice configuration, the binding energy is lower by ~ 2 eV than prior to subtraction.

The relative stability of the [110]-orientated electron bipolarons can be established by comparing the bipolaron state with two isolated small polarons which are formed upon dissociation. Corresponding embedded cluster calculations employing the simplified cluster representation shown in Fig. 3.17 indicate bipolaron stabilization energies of the order of 0.1 eV. Therefore gap levels referring to stable electron bipolarons are located close to the conduction-band states. Again, bipolarons turn out to be unstable within HF theory.

There are no indications in favour of [100]-orientated electron bipolarons. Obviously, the oxygen anion separating the two Ti^{3+} cations impedes the formation of these species. Thus, one could speculate about the formation of [100]-orientated electron bipolarons in the case of existing oxygen vacancies (see Sect. 3.3.4).

Although the preliminary investigations sketched above provide significant hints towards electron bipolaron formation, it will be necessary to re-examine the calculations at an improved level, i.e. by including the complete oxygen ligand coordination octahedra of both Ti cations.

3.3.4 Embedded Cluster Calculations of Oxygen Vacancies in $BaTiO_3$

The major source of oxygen vacancies (F^{++} centres or $V_O^{\bullet\bullet}$ in Kröger–Vink notation [205]) in oxide perovskites refers to charge compensations of acceptor-type impurity cations, which are always dissolved into these materials during

[4] Isolated conduction-band electrons are likely to become localized at Ti cations to form small polarons. This is due to lattice relaxation effects and to possible Anderson localization upon disorder. Their electronic ground state is $3d_{xy}$. Jahn–Teller distortion provides further stabilization.

crystal growth, with significant concentrations of the order of some hundreds ppm (see also Sect. 3.2). Thus, the same order of magnitude should be expected for F centres. Intrinsic defect reactions, on the other hand, are of only minor importance in this respect. Another effective source of oxygen vacancies relates to annealing treatments in reducing atmospheres. Here each created oxygen vacancy serves as a charge compensator of two electrons released to available electronic crystal states, i.e. to conduction band or donor-type defect states.

There are several reasons to investigate the properties of oxygen vacancies in oxide perovskites: oxygen vacancies represent donor-type defects which, in principle, are able to bind two electrons simultaneously. The trapping of one electron defines singly charged F centres (F^+), whereas two trapped electrons refer to neutral oxygen vacancies (F^0). Due to the estimated concentrations of such defects one should expect significant influences on the observed n-type conductivity and on the optical absorption properties of the materials. This is of particular interest towards a complete characterization of photorefractive material properties. In the present context it is important to study the favourable electronic states of electrons trapped at oxygen vacancies and to calculate the geometrical pattern of the resulting defect complexes.

Before discussing embedded cluster calculations, I review some of the recent experimental and theoretical investigations on oxygen vacancies.

The n-type conductivity of $BaTiO_3$ can be explained if oxygen vacancies are assumed to be completely ionized at elevated temperatures $T > 600$ °C. This result indicates the existence of shallow gap levels related to oxygen vacancies. However, up to this day there is no proven structural model of these centres. Previous shell model simulations [158, 206, 207] suggested electronic ionization energies of about 0.1 eV. This agrees with recent experimental ionization energies (i.e. 0.2–0.3 eV [208]) which were derived from conductivity measurements. The shell model estimate rests on the representation of F^+ centres as symmetry-broken Ti^{3+}–$V_O^{\bullet\bullet}$ defect complexes with the electron being localized at exactly one of the two Ti cations next to the vacancy; the electronic ionization energy has been obtained by combining shell model defect energies with free-ion ionization potentials. Additional support in favour of this structural model seems to derive from empirical Green's function investigations [57, 60, 62, 209], which were briefly reviewed in the introduction of this section 3.3. However, it is emphasized again that these latter results may be considerably modified upon changing the (empirical) model parametrizations. Moreover, these calculations do not include Ti 4s and 4p orbitals. Indeed, earlier discrete variational (DV) X_α cluster calculations [210] emphasized the importance of such additional orbitals accomplishing pronounced hybridizations between Ti $e_g(3d_{3z^2-r^2})$[5] and excited 4s and 4p orbitals. Another substantial point refers to the inclusion of vacancy-centred basis functions, which allow the transfer of electron density onto the

[5] The z-axis corresponds to the main axis of the defect complex.

vacant oxygen site. This obvious degree of freedom has been neglected completely in the earlier Green's function simulations, but has been accounted for in the present embedded cluster calculations. The implications of these extensions, leading to substantial modifications of the structural model of oxygen vacancies, will be discussed below.

Very recently Scharfschwerdt et al. [211] published the details of careful electron spin resonance investigations on paramagnetic defect centres observable in reduced $BaTiO_3$. The authors suggest interpretations in terms of F^+ centres assumed to consist of symmetry-broken Ti^{3+} $V_O^{\bullet\bullet}$ defect complexes, as discussed above. Anticipating this interpretation, the investigations demonstrate the existence of two classes of F^+ centres, i.e. axial Ti^{3+}–$V_O^{\bullet\bullet}$ complexes orientated along $\langle 001 \rangle$ directions and non-axial centres, of which the g-tensor is moderately tilted against $\langle 001 \rangle$. For group-theoretical reasons the electronic ground state could be identified in both cases as Ti $t_{2g}(3d_{xy})$ which is orientated perpendicularly to [001]. In the case of non-axial centres the observed tilting has been ascribed to perturbing alkali acceptor cations incorporated at nearby Ba sites. Interestingly, only non-axial centres are observed if the $BaTiO_3$ samples are alkali-contaminated. This indicates the importance of acceptor associations. The axial centres have been assigned to isolated oxygen vacancies.

The present embedded cluster calculations are used to provide independent arguments concerning the microscopic structure of oxygen vacancies. The calculations are based on the extended 34 atom cluster $Ti_2O_{10}Ba_{12}Ti_{10}$ (Fig. 3.18) which correctly models the local mirror symmetry of the considered oxygen vacancy. The embedding lattice is simulated using a shell model representation. Further details of this methodology are delineated in Sect. 2.1.2. All simulations are based on the cubic phase of $BaTiO_3$. This simplification may be justified due to the observation that all possible ferroelectric distortions of the material are small compared with the usual defect-induced lattice relaxations.

The quantum mechanical description of the central defect cluster employs Gaussian-type basis functions with split valence quality for all titanium cations and oxygen anions inside the cluster; the oxygen basis set is augmented by polarizing d and diffuse p functions. Further, three s and three p-type basis functions with exponents 0.05, 0.1 and 1.0 have been implemented at the vacant oxygen site. Finally, bare effective core potentials (ECP) [21] are used to model the localizing ion-size effects of all Ba and Ti cations at the cluster boundary.

The *ab initio* level of the present calculations covers Hartree–Fock (HF) theory including MP2 corrections and density functional theory (DFT), the latter of which use the GGA-type "BLYP" exchange correlation functional.

As a starting model we consider the structure of isolated singly charged oxygen vacancies F^+. It is natural to assume that this model should display all the main features known experimentally. Figure 3.19 displays electron

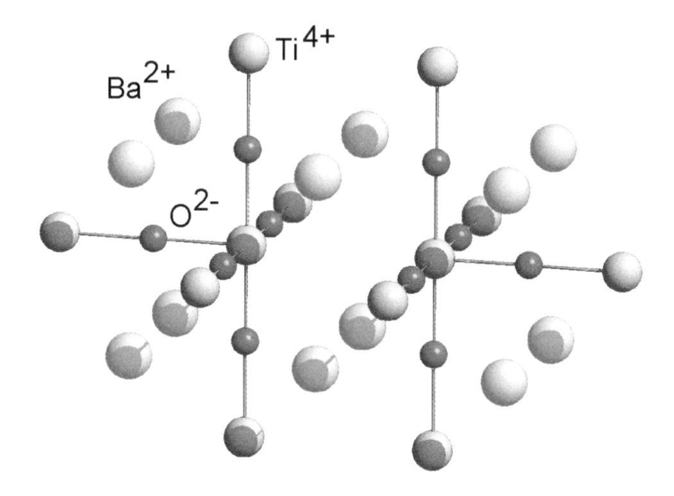

Fig. 3.18. Visualization of the extended $Ti_2O_{10}Ba_{12}Ti_{10}$ cluster. The oxygen vacancy is located at the centre of this cluster; all further ion positions in this picture correspond to the undistorted cubic lattice structure of $BaTiO_3$

density plots of a single electron trapped in Ti e_g or in t_{2g} states[6], alternatively. To avoid t_{2g}-nodes, density contributions are sampled above the yz-plane containing the central Ti–V_O–Ti complex. MP2 and DFT calculations consistently show that the e_g state is by about 1 eV more favourable than the shallow-gap t_{2g} state[7]; even within the first-order HF approximation the corresponding energy separation is 0.5 eV. Thus, seemingly different from experience, the calculations suggest a deep gap ground state level. Figure 3.19 indicates the occurrence of symmetry breaking only in the case of the t_{2g} state, whereas the e_g state refers to a highly symmetrical solution with the electron equally delocalized over the vacancy and both of its next Ti neighbours. Regarding the localized t_{2g} state, it is observed that lattice relaxation stabilizes the breaking of symmetry. The calculations also show that the energetic preference of e_g relates to hybridizations with excited Ti 4s and 4p orbitals and is further supported by the implementation of vacancy-centred orbitals. Only if we artificially omit these important degrees of freedom do we obtain a t_{2g}-type ground state – as in the empirical Green's function calculations. Note that the e_g state situation resembles proper F^+ centres, as are known to exist in ionic alkali halides. In these centres electrons are localized in s-type orbitals centred at the vacant anion site. In $BaTiO_3$ we observe a

[6] The notation is adapted to the titanium site symmetry.

[7] The ionization energy of this state amounts to a few tenths of an eV. It does not couple significantly with the environment, and the oxygen vacancy remains a perturbation.

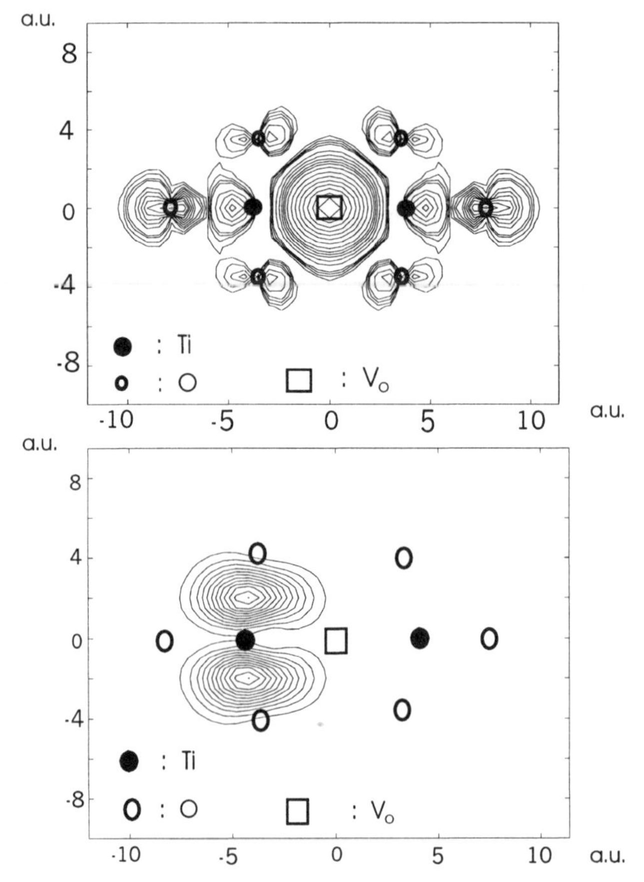

Fig. 3.19. Visualization of the density of an electron trapped at an oxygen vacancy (upper picture: localization in an e_g state; lower picture: localization in a t_{2g} state). The results are obtained from ΔSCF calculations using DFT-BLYP

significant mixing between s-type vacancy orbitals and the neighbouring Ti $e_g(3d_{3z^2-r^2})$ states, which refers to the semi-ionic nature of this material.

The stability of the present embedded cluster results, which are in good agreement with the earlier DV-X_α cluster calculations of Tsukada et al. [210], is impressive; in fact, they are independent of the inclusion of lattice relaxations. Moreover, this important result is not an artificial effect due to cutting off the electronic structure beyond the cluster surface within the embedded cluster representation, but it bears physical reality! Indeed, independent FLMTO supercell calculations of F centres provide additional strong support [212]. Corresponding results suggest an energy separation of the order of almost 1 eV in favour of e_g. Thus, from a theoretical point of view, we may conclude that properties of singly charged isolated oxygen vacancies are completely different from what is known experimentally. They cannot explain the main features of corresponding ESR centres observed in reduced $BaTiO_3$. It is finally recalled, that electrons trapped in e_g-type orbitals would lead to significantly different ESR signals compared with the t_{2g} situation[8].

It is also of certain interest to consider the geometrical relaxation pattern of F^+ centres. Generally, embedded cluster calculations show that nearby oxygen anions become attracted towards the oxygen vacancy, whereas cations are repelled. In particular, neighbouring Ti^{3+} cations are repelled by about 0.21 Å, if tacitly assumed in t_{2g} states. However, this displacement vanishes almost to zero if these cations are taken in their e_g-type ground states.

We are now in a position to speculate on model modifications towards consistent interpretations of the experimental data. What is the reason for the absence of ESR signals from isolated e_g-type F^+ centres in $BaTiO_3$?

As a first guess we could assume that all possibly existing isolated oxygen vacancies are used to form diamagnetic F^0 centres by trapping a second electron. This situation would correspond to a [100]-orientated electron bipolaron. The two spin-paired electrons would localize in e_g-type molecular orbitals. Though being ESR silent, this hypothetical bipolaron should contribute specific observable optical absorption bands, which, however, have not been reported so far. Moreover, we should expect the formation of paramagnetic e_g-type F^+ centres upon modest heating of the crystal samples. Further, from theoretical simulations there are no unambiguous indications in favour of these centres. Probably these centres would dissociate spontaneously according to

$$F^0 + F^{++} \longrightarrow 2F^+ . \tag{3.12}$$

In this sense, F^0 centres are expected to be of positive U type, with U being of the order of a few tenths of an eV. In summary, a bipolaron-based explanation seems to be unlikely.

Next, we discard the existence of isolated oxygen vacancies in favour of acceptor-associated defect centres. Suitable acceptors not only reduce the symmetry to non-axiality, but could also suppress the formation of deep gap e_g-type levels, leaving only the shallow gap t_{2g} states. In this context, acceptor cations are no longer considered as weak perturbations, but should strongly determine the electronic properties of oxygen vacancies. We recall that previous shell model simulations (see Sect. 3.2) predict the association of acceptor impurities at oxygen vacancies to be highly favourable. Even two acceptors may be bound simultaneously with equal affinity. A complete saturation of oxygen vacancies with acceptor cations has also been postulated in fluorite-structured MO_2 oxides [158, 213]. For a majority of trivalent cations this would imply the association of two acceptor impurities simultaneously. Further indications towards acceptor association derive from ESR investigations in $BaTiO_3$ (see [211]).

---

[8] For example, due to vanishing coupling matrix elements we should expect for the Zeeman coupling g-value $g_{||} \simeq g_e$ (for $z \parallel \boldsymbol{B}$), i.e. the free-electron g-value. But this is decisively different from the observed value.

Embedded cluster calculations suggest that Ba and Ti site acceptor cations associated one at a time are not effective in suppressing e_g levels. This is due to acceptor-induced lattice relaxations, which effectively screen the misfit charge of the associated impurities: before lattice equilibration due to the additional acceptor, the negative acceptor charge of di- and trivalent Ti site acceptors suffices to suppress the e_g state being lowest. This result, however, is completely reversed in favour of the e_g state if acceptor-induced lattice equilibration is allowed to take place. Also, it is just this screening of an acceptor misfit charge that causes the binding of a second acceptor to be as favourable as binding of the first acceptor. This can be inferred from previous shell model simulations described in Sect. 3.2. Finally, Ba site acceptors are completely unable to suppress e_g even before lattice relaxation. Thus, as a final model, we may conclude that only the simultaneous association of two Ti site acceptors is able to inhibit the formation of deep level e_g states. In this hypothetical situation conduction band electrons can localize only at second-neighbour Ti cations; the acceptor–vacancy complex remains a weak perturbation with respect to the cubic site symmetry of the second-neighbour Ti cation, and we thus expect a shallow gap t_{2g} ground state in agreement with presently available experimental information. A crucial experiment of this suggestive scenario consists of ESR measurements on extremely pure (i.e. acceptor-free) reduced $BaTiO_3$ crystals. Only then should we expect ESR signals of F^+ centres with electrons trapped in e_g orbitals. Additional conductivity measurements would complete the proof.

4. Potassium Niobate and Potassium Tantalate

These two oxides belong to the $A^{1+}B^{5+}O_3^{2-}$ family of perovskite-structured crystals. Whereas $KNbO_3$ shows the same sequence of FE-PT as $BaTiO_3$ (see Sect. 3.1), $KTaO_3$ remains cubic down to lowest temperatures. Its dielectric susceptibility, however, rises to about 4500 at 4K [214], suggesting that this oxide is close to a ferroelectric instability.

Recent band structure calculations [48, 215, 216] confirm the generally anticipated interpretation that the valence band (VB) has a mainly oxygen 2p character, whereas the conduction band (CB) essentially consists of Nb 4d (Ta 5d) orbitals. The Nb(Ta)–O hybridization leads to significant d contributions in the lower VB region. The calculations further show that both oxides possess an indirect band gap (VB–M versus CB–Γ) corresponding to $E_{VB}(M) - E_{VB}(\Gamma) \approx 0.7$ eV. Although underestimated by 40–50%, the LDA gap values for both perovskites yield the correct ordering, i.e. $E_g(KTaO_3) > E_g(KNbO_3)$. This difference can be qualitatively understood on the basis of the fifth free-ion ionization potentials of niobium (≈ 50 eV) and tantalum (≈ 48 eV). It may be argued that $KTaO_3$ reacts slightly more ionically than $KNbO_3$. Shell model and quantum cluster calculations support this interpretation (see Sect. 4.1), but also the observed saturation of the bonding peak in the $KTaO_3$-VB-DOS upon lattice distortions [48] can be interpreted due to stronger ionicity. In $KNbO_3$, on the other hand, the bonding peak increases upon ferroelectric lattice distortions. The covalency differences between $KNbO_3$ and $KTaO_3$ seem to be responsible for the different ferroelectric behaviour of both materials (see also Sect. 4.1).

$KTaO_3$ and $KNbO_3$ can be mixed resulting in the formation of the important electrooptic material KTN ($KTa_{1-x}Nb_xO_3$, $0 < x < 1$). Below the critical concentration $x_c < 0.008$ KTN remains cubic, but becomes polar (possibly ferroelectric) above this value. Unambiguously ferroelectricity has been established to occur for $x \geq 0.05$ (see [217] for a review). The electrooptic material properties of KTN may be tailored by adjusting x. As for other electrooptic oxides the photorefractive properties depend on the presence of suitable defect centres. Upon their natural abundance iron cations represent the dominating impurity defects in this respect. Section 4.1 summarizes the shell model simulations of both perovskites, the defect chemistry is reviewed in Sect. 4.1.2. The charge carrier transport taking place in the course of photorefractive processes in $KNbO_3$ is maintained by holes (VB) and electrons (CB), but the sensitivity of the material becomes enhanced upon electrochemical reduction making electrons the dominant carriers [218].

The introduction of Nb^{5+}_{Ta} defects in $KTaO_3$ leading to the formation of KTN represents an important example of a polarizing defect. Further defects of this type are given by Li^{+}_{K} and Na^{+}_{K} (see [219, 220] for a review of the field). The off-centre displacement of these impurity cations is responsible for their dipole character. In $KTaO_3$ polar phases appear upon doping with corresponding impurities [221]. Also (permanent) dipole defect centres like M^{n+}–$V^{\bullet\bullet}_{O}$ (M being any transition metal cation, i.e. either extrinsic or intrinsic) and $M^{n+}_{K}-O^{ll}_{I}$ may be considered as polarizing defects. The corresponding defect centres related to Fe^{3+} show characteristic axial ESR spectra which may be analyzed geometrically using the superposition model of Newman and Urban (see Sect. 4.2). However, unambiguous results are only obtained if data from atomistic simulations are available. Corresponding investigations are discussed in Sect. 4.2.2. The $Fe^{3+}_{K} - O^{ll}_{I}$ has been recently observed by ODMR (optically detected magnetic resonance) techniques [222]. The defect centre can be aligned using linearly polarized laser light.

Recent second harmonic generation (SHG) studies on $KTaO_3$ [223] provided evidence for approximately 10^{18} cm^{-3} dipole centres, for which the concentration rises upon reduction and decreases by oxidizing processes. Therefore the dipole centres might be identified with defect complexes containing oxygen vacancies.

Section 4.2 is devoted to a general discussion of polarizing defects.

4.1 Shell Model Simulations in $KNbO_3$ and $KTaO_3$

In this section we consider shell model simulations of perfect and defective $KNbO_3$ and $KTaO_3$. The potential parameters for both oxides have been mostly derived empirically. In addition, particular emphasis will be paid to the *ab initio* hydrostatic Nb^{5+}... O^{2-} short-range potential which has been discussed in Sect. 2.2.2.

4.1.1 Shell Model Parameters and Perfect Lattice Simulations

$KTaO_3$. The shell model parameters used in this study (A, ϱ and C for each anion–cation pair and Y, k for each ion) are determined by empirical fitting procedures. Only the oxygen–oxygen short-range interaction parameters A and ϱ are taken from earlier Hartree–Fock cluster calculations [224]. It is recalled that in the empirical approach the unknown parameters are treated as variables and are adjusted to achieve best agreement between calculated perfect crystal properties and the corresponding experimental data (see Sect. 2.2 for details).

For $KTaO_3$ a relaxed fitting procedure is employed, which involves the calculation of properties to be carried out for a relaxed lattice (see Sect. 2.2.2). This proves necessary for $KTaO_3$, since minute changes of ion positions, which occur when residual strains on the structure are released after a conventional "unrelaxed" fitting, result in large changes of dielectric constants.

Structural properties as well as elastic and dielectric constants of $KTaO_3$ are used in the fitting procedure to obtain the potential parameters compiled in Table 4.1.

Table 4.1. Empirically derived shell model potential parameters for $KTaO_3$

Interactions	Short-range potential parameters		
	A/eV	ϱ/Å	C/eVÅ^6
O^{2-}... O^{2-}	22764.3	0.149	27.627
Ta^{5+}... O^{2-}	1315.572	0.36905	0.0
K^+... O^{2-}	523.156	0.34356	0.0

Ion	Shell parameters	
	Y/e	k/eVÅ^{-2}
O^{2-}	−2.75823	30.211
Ta^{5+}	−4.596	5916.770
K^+	–	–

In Table 4.2 experimental properties are compared with the respective calculated ones.

$KNbO_3$. The empirical shell model potential parameters used to model $KNbO_3$ are transferred from simulation studies of KNb_3O_8. The parameter set has been obtained by fitting only to structural data available for this oxide [12]. As for $KTaO_3$ the inter-oxygen short-range potential is retained from earlier HF calculations [224].

Both the empirical as well as the *ab initio* Nb^{5+}... O^{2-} short-range potential generate, when used in conjunction with the remaining empirical model parameters, the cubic phase of $KNbO_3$, but, at present, there is no satisfactory account of ferroelectric distortions. With the *ab initio* potential the cubic lattice constant of $KNbO_3$ is overestimated by only 0.47% (but 1.67% employing the empirical potential). The accurate description of ferroelectric phases would demand to go beyond the pair potential and dipole polarizable ion approximation. This task requires further that *ab initio* potentials should be derived from supercell calculations in order to catch some of the cooperative phenomena governing ferroelectricity. Application of the hydrostatic

Table 4.2. Comparison of calculated and observed properties of $KTaO_3$: $C_{\mu\nu}$= elastic constants, ε_0= static and ε_∞= high-frequency dielectric constant, E_{lat}= lattice energy. All properties except for E_{lat} have been included in the fitting procedure

Property		Theoretical	Experimental	
C_{11}	$(10^{10}Nm^{-2})$	39.88	39.36	[a]
C_{12}	$(10^{10}Nm^{-2})$	10.57	–	
C_{44}	$(10^{10}Nm^{-2})$	10.90	10.71	[a]
ε_0		244.08	243.00	[b]
ε_∞		4.426	4.592	[c]
E_{lat}	(eV)	−174.73	−180.98	[d]

[a]H. H. Barrett. *Phys. Lett.*, 26A:217, 1968
[b]S. H. Wemple. *Phys. Rev.*, 137:1575, 1965
[c]Y. Fujii and T. Sakudo. *J. Phys. Soc. Japan*, 41:889, 1976
[d]From Born–Haber cycle

potential generation method (see Sect. 2.2.2) in combination with additional symmetry-breaking (i.e. ferroelectric) lattice distortions could result in reasonable interionic potential and ionic shell parameters (see also Sect. 2.2). In the present context the main interest is devoted to the calculation of defect formation energies. Therefore it is sufficient to employ the actual potential parameters generating the cubic crystal phase, because energetic differences between the various crystallographic phases are very small compared with commonly obtained defect energies. Table 4.3 compiles the shell model parameter sets which are used in the subsequent simulations. A reasoning for the employed (empirical) shell parameters in conjunction with the *ab initio* Nb^{5+}... O^{2-} short-range potential will be given below.

Table 4.4 compares some macroscopic crystal properties of $KNbO_3$ calculated using the *ab initio* and empirical Nb^{5+}... O^{2-} potentials. Note that at this stage we do not include any shell parameters for Nb^{5+} and O^{2-} ions in combination with the *ab initio* interaction. It is recalled that the hydrostatic procedure does not allow us to derive these parameters (below, however, we shall discuss *ad hoc* introduced ion polarizabilities and their influence on calculated crystal properties).

By inspection of Table 4.4 it is observed that both potential sets provide a satisfactory description of the structural and elastic properties of cubic $KNbO_3$. In fact, these entities are determined by interionic potentials and not by shell parameters. The observed deviations between calculated and experimental elastic properties are a consequence of the simplifying restrictions imposed *a priori* upon the potential model, e.g. pair potentials in central field approximation and dipole polarizable ions. Corresponding deviations even occur in the cubic and ionically bonded MgO. In this case many-body

Table 4.3. Short-range potential and shell parameters as used in subsequent shell model simulations. Because of their pronounced ionicity, potassium ions are modelled as rigid ions throughout the work. In the case of the *ab initio* Nb...O potential a range of different niobium spring constants are used in order to model different static dielectric constants (see text)

Short-range potential parameters				
Interaction type		A (eV)	ϱ (Å)	C (eVÅ^6)
O^{2-}... O^{2-}		22746.30	0.14900	27.88
Nb^{5+}... O^{2-}	(*ab initio*)	1333.44	0.36404	—
	(empirical)	1796.30	0.34598	—
K^{+}... O^{2-}		1000.30	0.36198	—

Shell parameters		
Ion	$Y(\|e\|)$	k (eVÅ^{-2})
Nb^{5+} (*ab initio*)	−4.496	2100 – 2500
Nb^{5+} (empirical)	−4.496	1358.58
O^{2-}	−2.811	103.07
K^{+}	—	—

Table 4.4. Calculated and experimental macroscopic crystal properties of $KNbO_3$. In the case of the *ab initio* Nb...O potential a rigid ion potential model is used at this stage. Except for the dielectric behaviour, however, all other properties turn out to be insensitive to the implementation of ionic polarizabilities

Property	Calculated		Observed[a,b]
	ab initio	Empirical	
C_{11} (10^{10} Nm^{-2})	39.788	43.349	25.5
C_{12} (10^{10} Nm^{-2})	14.178	13.114	8.0
C_{44} (10^{10} Nm^{-2})	14.178	13.114	9.0
ε_0	4.884	23.789	—
ε_∞	1.000	1.806	—
Lattice constant (Å)	4.026	4.073	4.007

[a] A. C. Nunes, J. D. Axe, and G. Shirane. *Ferroelectrics*, 2:291, 1971
[b] E. Wiesendanger. *Ferroelectrics*, 6:263, 1974

corrections turned out to be sufficient in order to correct for the deficiencies immanent in ordinary shell model simulations (see [106], for example). It is finally emphasized that the potential models for $KNbO_3$ (different from $KTaO_3$) do not involve any fitting to macroscopic material properties.

In contrast to the structural and elastic properties the dielectric behaviour is very sensitive to the choice of shell parameters, as is shown in Table 4.5. In particular, the static dielectric constant assumes each arbitrary positive value (above the rigid ion value) by changing the shell parameters appropriately. Figure 4.1 displays the static dielectric constant as a function of the niobium harmonic spring constant.

A similar plot may be generated if the oxygen harmonic spring constant is varied instead. However, it seems instructive that appropriately changing the polarizability of niobium ions, which are significantly less polarizable than oxygen ions ($\alpha_O/\alpha_{Nb} \approx 9$ according to the present shell parameters), leads to a behaviour which resembles a ferroelectric instability of the modelled $KNbO_3$. Indeed, upon using $k_{Nb} = 1900$ eVÅ^{-2} (which is clearly below the singularity in Fig. 4.1) the cubic structure becomes unstable (due to negative dielectric constants). Perfect lattice relaxations suggest that a rhombohedrally distorted structure is by 0.03 eV per formula unit more favourable in this situation. Most importantly, the rhombohedral structure reveals a [111] niobium off-centre displacement of about 0.19 Å, but unfortunately the distortion of the oxygen octahedron does not comply satisfactorily with observation.

Table 4.5. Calculated dielectric behaviour of $KNbO_3$ as a function of ionic polarizabilities using the *ab initio* hydrostatic Nb...O potential

Variation of k_{Nb} (eVÅ^{-2}) ($k_O = 103$ eVÅ^{-2})	ε_0	ε_∞
∞	7.703	1.680
2500.0	31.323	1.775
2200.0	100.510	1.790
2120.0	431.690	1.795
2100.0	3937.908	1.796
Variation of k_O (eVÅ^{-2}) ($k_{Nb} = 2500$ eVÅ^{-2})		
90.0	71.649	1.775
85.0	179.332	1.910
83.0	517.565	2.005
82.0	> 10000	2.021

In order to derive a physically reasonable shell parameter set one may suggest the following empirically based model. First, we essentially retain the empirical oxygen shell parameters (i.e. $Y_O=-2.811\ |e|$, $k_O=103.07$ eVÅ^{-2}) from the successful simulation studies of the related material KNb_3O_8 [12]. Second, potassium ions are simulated as being unpolarizable because of their pronounced ionic nature. Third, with respect to Nb^{5+} we only consider variations of the harmonic spring constant k_{Nb} while keeping the shell charge equal to the empirical value (i.e. $Y_{Nb}=-4.496\ |e|$). This procedure, although oversimplified, is sufficient to investigate the effects of ionic polarizabilities. Moreover, we choose k_{Nb} close to its instability value.

The qualitative justification of our model concerning ionic polarizabilities is as follows: since we know that $KNbO_3$ is unstable against ferroelectric phase transitions, we choose both oxygen- and niobium-related shell parameters (i.e. core–shell spring constants) to be close to their respective instability

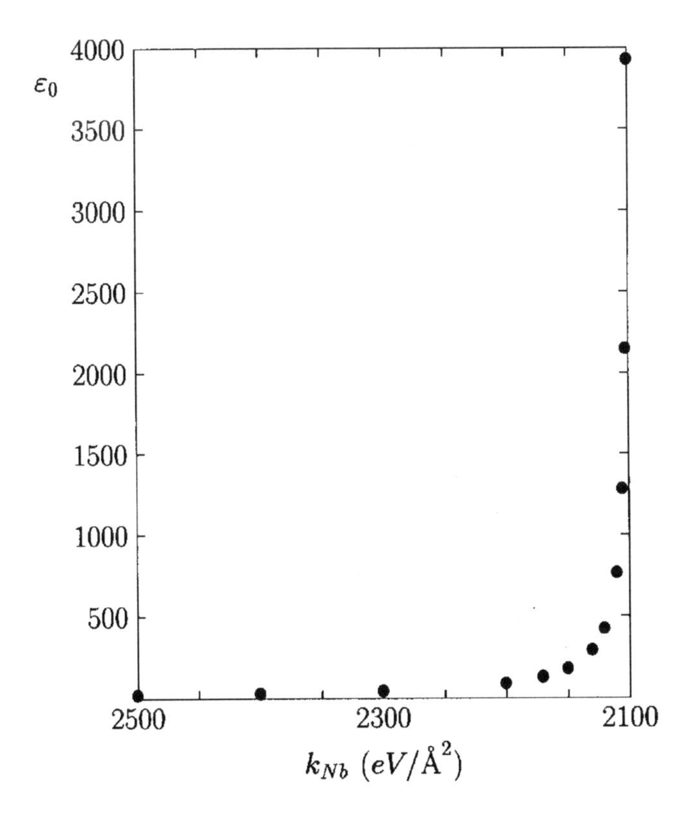

Fig. 4.1. The static dielectric constant as a function of the niobium core–shell harmonic spring constant. The simulations employ the *ab initio* Nb...O short-range potential

region (see Table 4.5, Fig. 4.1). The importance of ionic polarizabilities with respect to a ferroelectric behaviour has already been stressed by Bilz et al. [138]. However, in contrast to the present model these authors ascribe the dominant contributions to the oxygen polarizability. Support in favour of this interpretation seems to be given by shell model simulations, since, in particular, the TO phonon softening may be reproduced in this way. On the other hand, *ab initio* electronic structure calculations based on the LAPW method [225, 226] strongly suggest that covalency in terms of hybridizations between O 2p and Nb 4d orbitals is even more important than the oxygen polarizability (see also Sect. 3.1). As a consequence, in ABO_3 perovskites the most pronounced changes of the electron density upon ferroelectric distortions are located at the B cations and not at the oxygen sites. In shell model simulations this behaviour can be most closely mimicked by varying the polarizability of the B cations.

A comparison of different but closely related materials ($KNbO_3$ and $KTaO_3$ in the present context) can provide additional support in favour of the present polarizability scenario. Whereas $KNbO_3$ undergoes ferroelectric phase transitions, these are inhibited in $KTaO_3$. Since the most obvious difference between both crystals, which have the same lattice structure with almost equal interionic separations, is provided by the different B cations, the differences in ferroelectric behaviour should be related to these cations. Indeed, by comparing the shell model parameter sets of $KNbO_3$ and $KTaO_3$ it is observed that, first, the Ta...O short-range interaction is slightly more repulsive than both the *ab initio* and the empirical Nb...O potentials and that, second, Nb ions are more polarizable than Ta ions, but oxygen anions are less polarizable in $KNbO_3$ than in $KTaO_3$. We find $\alpha_O/\alpha_{Ta} \approx 70$ upon inserting the appropriate shell parameters; thus, this ratio is significantly larger than in $KNbO_3$. These differences, which are probably not due to the ambiguities of shell model simulations, may be explained by assuming a stronger ionicity of the tantalum–oxygen bonds in comparison to the niobium–oxygen counterparts. Also, the free-ion ionization potentials of niobium (50 eV) and tantalum (48 eV) suggest this expected behaviour, of which the origin might be traced back to the lanthanide contraction of tantalum. A reduced covalent charge transfer in $KTaO_3$ crystals would cause the oxygen ions to remain more polarizable than in $KNbO_3$, whereas Nb ions, in turn, should be more polarizable than Ta ions. Covalent charge transfer from oxygen onto the octahedrally coordinated metal ion stabilizes the oxygen ions, which become less negative, and destabilizes the metal ion because of its increased electronic charge.

The covalency-related interpretation of shell model ionic polarizabilities is supported by *ab initio* cluster calculations employing cubic $BO_6K_8^+B_6^{5+}$ (B=Ta or Nb) clusters with the interionic separations chosen along the observed lattice constants (4.007 Å for $KNbO_3$ and 3.9885 Å for $KTaO_3$). The outer cations are modelled by bare effective core potentials [21]. Gaussian-type basis functions are used for the B cations [21] and for the oxygen anions [126]. The oxygen basis set is further augmented by polarizing d- and diffuse p-type functions. Table 4.6 compiles the results of Mulliken population analyses. The Δq charge differences defined in Table 4.6 indicate the direction of charge transfer differences. It is emphasized that ion charges derived from Mulliken population analyses should be taken qualitatively and not quantitatively. Possible ambiguities refer to the applied charge distribution methodology but also to the particularly employed basis functions [17]. However, all performed test calculations confirm the stability of the established ionicity differences. Moreover, additional simulations show that differences due to the deviating lattice constants may safely be neglected. The stronger ionicity of the Ta...O bond compared to Nb...O has also been stressed by Thomann [227]. It is further noted that the Nb polarizability ultimately allows these ions to move off the octahedral cubic lattice site in cubic $KTaO_3$ (Sect. 4.2.1).

Table 4.6. Mulliken population analyses for the niobium and tantalum clusters specified in the text. q_B and q_O denote the obtained Mulliken charges of B cations and oxygen anions, respectively. $\Delta q = q_{B(O)}$(Nb cluster) – $q_{B(O)}$(Ta cluster)

	UHF				DFT-BLYP		
	B = Ta	B = Nb	Δq		B = Ta	B = Nb	Δq
q_B	+2.800	+2.710	−0.090	q_B	+1.998	+1.882	−0.116
q_O	−1.638	−1.623	+0.015	q_O	−1.499	−1.480	+0.019

Final support in favour of the present Nb parametrization is given by simulations of the complex $LiNbO_3$ (see also Chap. 5) when using the *ab initio* Nb...O short-range potential combined with appropriate niobium shell parameters close to the instability region in Fig. 4.1. All remaining shell model parameters are retained from the earlier empirical model reported in [124]. Table 4.7 compiles calculated and observed properties of $LiNbO_3$. Differently from cubic $KNbO_3$, all material properties depend on the choice of potential and shell parameters. Table 4.7 suggests a very satisfactory agreement between calculated and observed properties. The only exception refers to C_{44} of which the slightly negatively calculated value indicates a structural instability. This defect should be removed before the complete parametrization can be used in defect simulations[1]. But the most important point I want to address in the present context refers to the observation that the niobium-

[1] In principle the necessary adjustment seems to be possible following the observation that this particular elastic constant depends sensitively on the shell and potential parameters. However, due to the expected small changes of calculated defect energies, such investigations will be of secondary importance. Therefore,

Table 4.7. Calculated and experimental crystal properties of $LiNbO_3$ using the *ab initio* Nb...O short-range potential and k_{Nb} = 2600 ÅeV^{-2} within the earlier empirical model

Property	Calculated	Observed[a]
C_{11} (10^{10} Nm^{-2})	23.2	20.3
C_{12} (10^{10} Nm^{-2})	11.8	5.3
C_{13} (10^{10} Nm^{-2})	6.6	7.5
C_{14} (10^{10} Nm^{-2})	3.9	0.9
C_{33} (10^{10} Nm^{-2})	19.0	24.5
C_{44} (10^{10} Nm^{-2})	−0.87	6.0
C_{66} (10^{10} Nm^{-2})	5.7	7.5
$\varepsilon_{0,11}$	72.6	84.1
$\varepsilon_{0,33}$	28.0	28.1
Lattice constant c (Å)	13.928	13.863

[a] R. S. Weiss and T. K. Gaylord. *Appl. Phys.*, A37:191, 1985

related shell parameters, which have been discussed above, allow the accurate reproduction of the observed static dielectric properties. On the other hand, this task becomes completely impossible by varying instead the oxygen shell parameters, and keeping the niobium cations more rigidly at the same time. These additional simulations of perfect $LiNbO_3$ emphasize the particular role of B cations, too.

4.1.2 Defect Chemistry of $KTaO_3$ and $KNbO_3$

Intrinsic defects. Table 4.8 summarizes defect formation energies for the most important intrinsic defects in $KNbO_3$ using the *ab initio* Nb^{5+}... O^{2-} potential for various choices of shell parameters. For comparison defect energies calculated on the basis of the empirical model are listed, too. We observe from Table 4.8 that corresponding defect energies vary by an amount between 5% and 20% as a function of ionic polarizabilities (note that ε_0 ranges between 8 and 4000 upon using the different shell parameters). As expected, defect energies decrease with increasing ionic polarizability. Moreover, the application of the *ab initio* Nb^{5+}... O^{2-} potential leads to smaller defect energies when compared to the empirical model. Most importantly, the defect chemical properties of $KNbO_3$, however, remain qualitatively unaltered upon these changes. Table 4.9 compiles the relevant defect formation energies concerning Frenkel and Schottky disorder, electron- and hole-like polarons and, finally, the electronic band gap.

all results reported in Chap. 5 are derived from the earlier empirical shell model parameter set [124].

Table 4.8. Formation energies (eV) of important intrinsic defects in $KNbO_3$. The last column displays results based on the empirical Nb...O potential parameters. t: thermal relaxation; o: optical relaxation. In the latter type of relaxation only the shells are allowed to be displaced. All spring constants are given in eVÅ^{-2}

	Ab initio Nb...O						Empir-
Defect	$k_O = 103.1$				$k_{Nb} = 2500.0$		ical
	k_{Nb}				k_O		
	2100.0	2300.0	2500.0	∞	90.0	83.0	
V_{Nb}^{5l}	119.6	121.9	123.6	133.3	121.4	120.0	134.8
$V_O^{\bullet\bullet}$	19.7	19.8	20.0	20.8	19.8	19.6	21.0
V_K^{l}	4.0	4.0	4.1	4.3	4.0	4.0	4.1
$Nb_K^{4\bullet}$	−109.7	−109.3	−109.0	−106.9	−109.5	−109.8	−109.4
$Nb_I^{5\bullet}$	−102.4	−101.3	−100.5	−94.2	−101.6	−102.3	−94.9
$K_I^{\bullet}$	3.9	4.0	4.0	4.4	4.0	3.9	5.9
O_I^{ll}	−7.3	−7.0	−68	−5.5	−7.2	−7.6	−5.9
Nb_{Nb}^{l} (t)	44.7	44.8	44.9	45.4	44.8	44.7	45.9
Nb_{Nb}^{l} (o)	48.4	48.5	48.5	48.6	48.2	87.0	48.7
$O_O^{\bullet}$ (t)	17.4	17.5	17.5	17.9	17.5	17.4	18.1
$O_O^{\bullet}$ (o)	23.2	23.2	23.2	23.3	23.1	23.0	23.2

In the case of Frenkel- and Schottky-type disorder we note that all reaction energies decrease in line with single defect energies if the *ab initio* potential is used with increasing ion polarizabilities. In particular, Nb^{5+} Frenkel disorder as well as Nb_2O_5- and $KNbO_3$-related Schottky-type defects take advantage of these changes. In spite of this, all defects remain energetically unfavourable to be created with higher concentrations in $KNbO_3$ crystals. It is remarked that the same qualitative conclusions may be drawn for $KTaO_3$ (for corresponding details see [228]). Obviously, all ABO_3 perovskites seem to behave similarly in that the major source for intrinsic defects refers to the incorporation of impurities (see Sect. 4.1.2 and Chap. 3).

Any shell model treatment of excessive electron or hole species implies a small-polaron model. The stability criterion favouring small polarons against large polarons or band-like species is given by

$$|\Delta E_B| > \frac{\Delta}{2}, \tag{4.1}$$

where ΔE_B is the small-polaron binding energy listed in Table 4.9 and Δ the appropriate band width. The small-polaron binding energy is defined as the difference between the optical and thermal polaron defect energies, where "optical" implies that only ionic shells are allowed to relax and "thermal" implies a full relaxation of cores and shells. The right-hand side of (4.1) rep-

Table 4.9. Calculated intrinsic defect chemical properties in $KNbO_3$. All spring constants are given in $eVÅ^{-2}$

	Ab initio Nb...O						Empirical Nb...O
Defect	$k_O = 103.1$				$k_{Nb} = 2500.0$		$k_O = 103$
type	k_{Nb}				k_O		k_{Nb}
	2100.0	2300.0	2500.0	∞	90.0	83.0	1358.58
Frenkel disorder, energy per defect (eV)							
K^+	3.9	4.0	4.0	4.3	4.0	4.0	5.0
Nb^{5+}	8.6	10.3	11.5	19.6	10.0	8.9	19.9
O^{2-}	6.2	6.4	6.6	7.7	6.3	6.0	7.6
Schottky-type disorder, energy per defect (eV)							
K_2O	1.8	1.8	1.9	2.4	1.8	1.7	2.3
Nb_2O_5	2.1	2.9	3.4	6.8	2.7	2.2	7.4
$KNbO_3$	1.8	2.4	2.8	5.3	2.2	1.9	5.7
Electron and hole polarons, $\Delta E_B = E_{optical} - E_{thermal}$ (eV)							
Nb^{4+}_{Nb}	3.7	3.6	3.6	3.2	3.4	3.3	2.8
O^{1-}_{O}	5.7	5.7	5.7	5.5	5.6	5.5	5.1
Electronic gap, charge transfer energy $O^{2-} \longrightarrow Nb^{5+}$ (eV)							
	2.9	3.0	3.1	3.9	2.9	2.8	4.7

resents the gain in delocalization energy in a band model which is competitive with small-polaron formation. In the case of electrons (Nb^{4+}_{Nb}) the conduction band width is Δ_{CB}=3 eV [229] and in the case of holes (O^{1-}_{O}) the width of the valence band Δ_{VB}=5 eV [229]. From this one can predict the formation of small electron and hole polarons to be favoured over band-like species for all potential models. Analogous results hold true for $KTaO_3$ [228]. It should be emphasized, however, that shell model simulations probably underestimate electronic contributions, leading to an overestimation of lattice relaxations at the same time. Therefore these calculations are naturally in favour of small polaron species. Deficiencies with respect to electronic terms might in particular affect the oxygen-related valence band contributions. Thus, one cannot fully exclude the stability of large hole-type polarons. On the other hand, the predicted existence of small electron polarons seems to be reliable, since the ESR spectra of Ta^{4+} and Nb^{4+} may be observed upon electrochemical reduction of the respective oxides. Moreover, the calculated hopping activation energy of small electron polarons in $KTaO_3$ (0.57 eV [228]) compares favourably with reported experimental data (0.37 eV [230] and 0.73 eV [231]).

Finally, Table 4.9 shows the calculated electron transfer energies from oxygen onto niobium. Since the oxygen 2p states define the valence band and niobium 4d states the conduction band, we may interpret the charge transfer energy to give the electronic band gap. The gap energies become very close to the experimentally reported value (3.1 eV [232]) upon using the *ab initio* potential in combination with niobium polarizabilities in the specified region above the expected "phase transition" (see Fig. 4.1).

Impurities. The discussion in the preceding subsection suggests that the majority of intrinsic defects is generated during the incorporation of impurity cations.

The subsequent description gives a brief review of the extrinsic defect chemistry of $KTaO_3$ and $KNbO_3$, for details the reader is referred to the references [207, 228]. The following reactions represent favourable solution modes for divalent cations in $KTaO_3$:

$$\mathrm{MO} +' \mathrm{KTaO_3}' \longrightarrow \frac{1}{2}M_K^{\bullet} + \frac{1}{2}M_{Ta}^{3l} + \frac{1}{2}V_O^{\bullet\bullet} + \frac{1}{2}\mathrm{KTaO_3} \tag{4.2}$$

$$\mathrm{MO} +' \mathrm{KTaO_3}' \longrightarrow \frac{3}{4}M_K^{\bullet} + \frac{1}{4}M_{Ta}^{3l} + \frac{1}{4}\mathrm{K_2O} + \frac{1}{4}\mathrm{KTaO_3} \tag{4.3}$$

$$\mathrm{MO} +' \mathrm{KTaO_3}' \longrightarrow M_K^{\bullet} + \frac{1}{2}K_{Ta}^{4l} + \frac{1}{2}V_O^{\bullet\bullet} + \frac{1}{2}\mathrm{KTaO_3} \tag{4.4}$$

The favoured solution reactions are given by the first two equations representing (partial) self-compensation modes. Self-compensation-type reactions represent the most favourable solution modes for all impurity cations M^{n+} having charge states $+2 \leq n \leq +4$. It is recalled that in $BaTiO_3$, in contrast, self-compensation is mostly restricted to trivalent cations.

Large cations (e.g. Ba^{2+}, Sr^{2+}, Ca^{2+}) prefer reaction (4.3), leading to a relative diminution of unfavourable Ta site defects. In all cases studied so far the third solution mode is less favourable, but the energy difference between (4.2) and (4.3) may become very small. Whereas trivalent cations prefer complete self-compensation, i.e.

$$\mathrm{M_2O_3} +' \mathrm{KTaO_3}' \longrightarrow \mathrm{M_K^{\bullet\bullet}} + \mathrm{M_{Ta}^{ll}} + \mathrm{KTaO_3}\,, \tag{4.5}$$

partial self-compensation involving the formation of potassium vacancies becomes equally favourable to complete self-compensation in the case of tetravalent impurity cations:

$$\mathrm{MO_2} +' \mathrm{KTaO_3}' \longrightarrow \frac{1}{3}M_K^{3\bullet} + \frac{2}{3}M_{Ta}^{l} + \frac{1}{3}\mathrm{V_K^{l}} + \frac{2}{3}\mathrm{KTaO_3} \tag{4.6}$$

$$\mathrm{MO_2} +' \mathrm{KTaO_3}' \longrightarrow \frac{1}{4}M_K^{3\bullet} + \frac{3}{4}M_{Ta}^{l} + \frac{1}{4}\mathrm{KTaO_3} + \frac{1}{4}\mathrm{Ta_2O_5}\,. \tag{4.7}$$

The defect chemical scenario is very similar in $KNbO_3$; however, there are additional competitive solution modes predicting impurity cations to completely occupy K sites. This is accomplished by electronic charge compensation, which reads for divalent cations:

$$MO +' KNbO_3{}' \longrightarrow M_K^{\bullet} + e^l + \frac{1}{2}K_2O + \frac{1}{4}O_2(g)\,, \tag{4.8}$$

where e^l denotes a small Nb_{Nb}^{4+}-polaron. However, referring to the discussion in Sect. 3.2, it is recalled that the use of free-ion ionization potentials generally overestimates the importance of electronic small-polaron compensations. In $KTaO_3$, on the other hand, this small-polaron compensation is unfavourable even due to the fifth free-ion ionization potential of tantalum. But at least in $KNbO_3$ one may expect that the existence of defect-induced gap levels could enhance the importance of electronic compensation mechanisms (see also Sect. 3.2).

In summary, the simulations of $KNbO_3$ and $KTaO_3$ indicate that most impurity cations would enter both cation sites, but in $KNbO_3$ the substitution of K ions could become predominant due to (defect-aided) electronic compensation. Recent channelling experiments are in support of this expectation [233]. In comparison to $BaTiO_3$, self-compensation of impurity cations is much more pronounced in the $A^{1+}B^{5+}O_3$ perovskites. For these materials it has been argued that the observed space charge limiting effects reducing the photorefractive efficiency could be related to impurity self-compensations [133].

We conclude our considerations by investigating the defect -chemical origin of the dipole defect centres $Fe_K^{3+} - O_I$ and $Fe_{Ta/Nb}^{3+} - V_O^{\bullet\bullet}$, which were mentioned in the introduction to this chapter. Whereas the latter defect complex may be readily explained due to existing oxygen vacancies (resulting from impurity solutions or electrochemical reduction), the questions now arise how the $Fe_K^{3+} - O_I$ centre can be created or whether O_I can be present in the crystal, because in general O_I ions are found to be energetically unfavourable in oxide perovskites.

Oxygen Frenkel defect pairs are not likely to be present, since their formation energy is 3.4 eV per defect. The reaction

$$Fe_2O_3 +' KTaO_3' \longrightarrow 2Fe_K^{3+} + 2O_I^{2-} + K_2O \tag{4.9}$$

requires 16.10 eV more than the self-compensation reaction (4.5) and is therefore energetically unfeasible. Even when taking the binding energy of the complex

$$Fe_K^{3+} + O_I^{2-} \longrightarrow Fe_K^{3+} - O_I^{2-} \tag{4.10}$$

of -3.53 eV into account the energy of (4.9) remains too high.

As a solution to the problem one can suggest the following reaction, where self-compensation is accepted as the main incorporation mechanism for Fe^{3+}

$$Fe_K^{3+} + Fe_{Ta}^{3+} \longrightarrow Fe_K^{3+} - O_I^{2-} + Fe_{Ta}^{3+} - V_O^{\bullet\bullet}\,. \tag{4.11}$$

The energy for (4.11) is $+1.35$ eV per oxygen defect complex. This "induced oxygen Frenkel defect"-type reaction reduces the energy to create oxygen interstitials by more than 2 eV compared with the simple oxygen Frenkel reaction. The energy is sufficiently low for appreciable concentrations of these species to be generated at high temperatures at which the crystals are grown. The trapped interstitials and vacancies will be immobile at low temperatures, hence their recombination is prevented.

Equation (4.11) also explains the simultaneous appearance of the centres $Fe_K^{3+} - O_I^{2-}$ and $Fe_{Ta}^{3+} - V_O^{\bullet\bullet}$. By combining defect formation energies from shell model calculations with the relevant ionization potentials it is found that the energy of the following charge transfer reaction (4.12) is +0.58 eV:

$$Fe_K^{3+} - O_I^{2-} \longrightarrow Fe_K^{2+} - O_I^{1-} . \tag{4.12}$$

This suggests that the defect centre $Fe_K^{3+} - O_I^{2-}$ is stable with respect to the indicated charge redistribution. However, the small value of the energy for (4.12) indicates that pronounced covalency should be present at this defect centre. In addition the present shell model simulations indicate that there is no appreciable energy barrier upon 90° reorientation of the defect complex. This result complies with the observed optical alignment of these axial iron centres [222]. The analogous centre in $KNbO_3$ is unstable (−0.90 eV for (4.12)) in agreement with its absence in ESR measurements in $KNbO_3$.

It is important to note that the above results are valid for non-reduced $KTaO_3$ crystals. Usual growth conditions, however, imply some degree of reduction of the material:

$$'KTaO_3' \longrightarrow V_O^{\bullet\bullet} + 2e^{l} + \frac{1}{2}O_2 . \tag{4.13}$$

The energy for (4.13) is 6.7 eV. Oxygen vacancies may be trapped, possibly at the Fe_{Ta}^{3+} centres, thereby providing another source for the observed $Fe_{Ta}^{3+} - V_O^{\bullet\bullet}$ centres. The electrons are trapped at low-lying electronic states in the gap, e.g. of Fe_K^{3+} centres. This reduces Fe^{3+} to Fe^{2+}. If no precautions are made to trap the electrons at other lower defect states, the Fe_K^{3+} centre becomes invisible in the ESR experiment. Experimentally the addition of Ti^{4+} ions [234] leads to observable Fe_K^{3+} and $Fe_K^{3+} - O_I^{2-}$ centres. The present shell model calculations show that Ti^{4+} is incorporated into $KTaO_3$ by the self-compensation-type reactions (4.6–4.7). The existing Ti_K^{4+} defects provide electronic states, which may be energetically more attractive for electrons than the Fe_K^{3+} states.

4.2 Polarizing Defect Centres

In this section we shall discuss the different types of existing polarizing defects. Regarding $KTaO_3$ in particular, these defects can facilitate the use of this material in electrooptic applications. A first characterization of these defects may be given as follows:

- Off-centre defects (Sect. 4.2.1). In most instances these defects refer to extrinsic cations substituting for potassium. It is emphasized that self-compensation-type solution modes provide a rich source of corresponding defects. The off-centre displacements of cations induce the creation of dipole moments.
- Formation of permanent dipoles due to defect aggregation (Sect. 4.2.2). The possible defect chemical origin of corresponding centres has been addressed in the preceding section.

Both types of defects show cooperative ordering features at sufficiently high doping levels (see [235, 236], for example). Some of these defect centres are related to trivalent iron impurities representing examples of paramagnetic S-state ions. The interpretation of electron spin resonance (ESR) data (i.e. zero-field-splitting parameters of the electronic ground state) of paramagnetic S-state ions in terms of the so-called superposition model (SPM) [237, 238] proves to be useful in order to understand the local geometrical crystal structure surrounding S-state cations like Fe^{3+}. On the basis of this model Siegel and Müller [239] could unambiguously show that isolated Fe^{3+} cations remain centred within their octahedral ligand coordination sphere. Thus, iron does not participate in the cooperative displacement of B cations against the oxygen sublattice, which to a considerable extent characterizes the ferroelectric crystal structure of ABO_3 perovskites. This result agrees with the observation that the Curie temperature decreases with increasing iron content [240, 241], if isolated Fe^{3+} ions represent the dominant iron-related defect species.

However, considering ligand coordination spheres, which differ from the simple octahedral cage, model-immanent ambiguities (see below) imply that SPM analyses have more qualitative predictive power rather than quantitative. But in this situation the combination of SPM analyses and additional shell model and/or embedded cluster calculations provides a powerful tool in order to single out all physically reasonable solutions. In particular, the considered defect-induced ion displacements may be chosen to comply with physically reasonable atomistic simulations. Correspondingly, combined investigations yield reliable characterizations of the geometrical microstructure of the Fe_B^{3+}–V_O and Fe_K^{3+}–O_I^{2-} defect complexes (see Sect. 4.2.2). It is emphasized that these defect centres have been subject to considerable misinterpretations prior to these refined investigations.

To prepare a common basis of understanding let us briefly summarize the fundamental ideas underlying the SPM. As was stated, the SPM principally allows us to give a geometrical interpretation of zero-field-splitting (ZFS) parameters occuring in the ESR spin Hamiltonian of S-state ions. Corresponding ions possess half-filled electronic subshells (e.g. Fe^{3+}, Mn^{2+} ($3d^5$) or Gd^{3+} ($4f^7$)) with maximal spin alignment. Covalency effects are supposed to provide the most important contributions to these ZFS parameters. Thus the SPM is restricted to scan the geometrical configuration with respect to the nearest ligand shell. A simple geometrical interpretation becomes possible upon the observation that all cation–ligand terms superpose linearly to sufficient approximation (see [238, 242] for explanations); e.g. for the axial ZFS parameter b_2^0 (=D) belonging to the spin Hamiltonian term $\propto (3S_z^2 - \boldsymbol{S}^2)$:

$$b_2^0 = \frac{1}{2} \sum_i \bar{b}_2(r_i)(3\cos^2(\theta_i) - 1) . \tag{4.14}$$

The sum runs over the nearest ligand shell and the polar coordinates refer to the main ESR axes with the S state ion at the origin. The radial function $\bar{b}_2(r)$ in (4.14) contains the impurity-specific information. Up to now there are no precise theoretical derivations of such radial functions, the main difficulty of which is related to the almost infinite number of contributions with comparable magnitude but with different sign [238]. It has been recognized that the radial function could be approximated using a Lennard–Jones-type function which collects all terms with equal sign in one power-law expression [238, 243]:

$$\bar{b}_2(r) = A\left(\frac{R_0}{r}\right)^M + B\left(\frac{R_0}{r}\right)^N . \tag{4.15}$$

Empirically these dependences are found to possess a minimum rather than a maximum. For 3d transition metal S-state cations a theoretical justification may be attempted by LCAO-based approaches to the ground state zero-field-splitting ESR parameter b_2^0 [242, 244]. These investigations indicate increasing $\bar{b}_2(r)$-values with increasing covalency, which is also expected upon reducing the relevant ion separations. Extensive work on Fe^{3+}-related $\bar{b}_2(r)$-parameters has been done by K. A. Müller and coworkers (e.g. [243, 245]). Upon uniaxial stress experiments performed on $MgO{:}Fe^{3+}$ and $SrTiO_3{:}Fe^{3+}$ the authors suggested the following parameters for (4.15): $A = -0.68$ cm^{-1}, $B = 0.27$ cm^{-1}, $M = 10$, $N = 13$ and $R_0 = 2.101$ Å. R_0 corresponds to the perfect lattice spacing in MgO. However, the solutions are not unique, because the only constraint can refer to definiteness close to R_0 (Fig. 4.2). Even the simplified monotonic inverse power expression,

$$\bar{b}_2(r) = \bar{b}_2(R_0)\left(\frac{R_0}{r}\right)^{t_2} , \tag{4.16}$$

approximates the more accurate Lennard–Jones-type function for r-values close to the perfect lattice spacing R_0 with $\bar{b}_2(R_0)$=-0.41 cm^{-1} and t_2=8. So far, almost all SPM investigations have been based on monotonic inverse power laws. But it should be noticed that such functions are likely to be unphysical for small ion separations. In addition, all $\bar{b}_2(r)$ functions may be subject to charge- and size-misfit effects and to coordination dependences which are able to modify the radial function parameters (see [246] and references cited therein).

Despite the scatter of possible radial functions one can draw useful qualitative conclusions from SPM analyses combined with appropriate atomistic simulations, if general Lennard–Jones-type radial functions are included

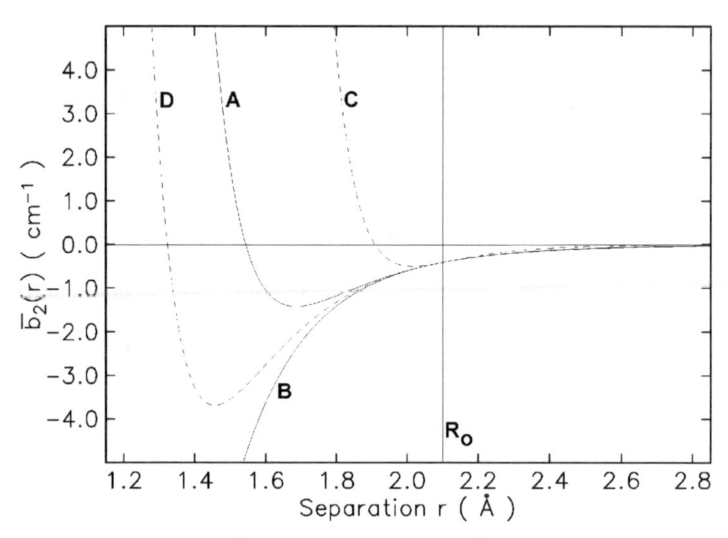

Fig. 4.2. Different $\bar{b}_2(r)$ functions. A, C, D: Lennard–Jones functions; B: inverse-power law. The different functions are chosen to agree close to the reference separation R_0. Curves A and B refer to the parameter sets specified in the text

rather than only inverse power laws. In this fashion artificial solutions can be avoided.

4.2.1 Dipole Formation Due to Off-Centre Displacements of Defects

Shell model and embedded cluster calculations can be used to study the tendency for isolated extrinsic cations to move off-centre, i.e. to show a significant displacement with respect to the perfect lattice site. The present discussion covers possible dependences upon the site of substitution and the impurity's size and charge state.

In the case of Ta site incorporation there are no off-centre displacements except for Si^{4+} and probably Nb^{5+}. These results can be understood in terms of ion-size and defect-charge effects. It is obvious that a large cation is strongly repelled by its oxygen ligands, which tend to keep the impurity cation in the centre of the oxygen cage. This argument also applies for defect charges lower than that of the substituted tantalum. However, ion-size effects seem to be of major importance as exemplified by Si^{4+} which shows an off-centre displacement of 0.87 Å corresponding to an energy gain of 2.95 eV. On the basis of this pronounced displacement the silicon ion is able effectively to reduce its coordination sphere from octahedral towards tetrahedral, which is the normal coordination of this cation. Most of the other cations seem to be well accommodated within an octahedral coordination sphere. In spite of the observed off-centre displacement this is also true for Nb^{5+}, because the absolute displacement is less than about 0.1 Å (see Sect. 4.2.1).

The observation that a pronounced off-centre displacement can be interpreted in terms of a tendency of cations to effectively change their coordination spheres immediately leads one to expect significant off-centre displacements in the case of K site incorporation. It is recalled from Sect. 4.1.2 that most cations can enter the potassium site by means of self-compensation-type solution modes. All cations investigated so far (apart from Rb^+ and Cs^+) are significantly smaller than the substituted potassium, a factor which favours reduction in the coordination number. This effect is amplified by the relative defect charges, since all higher valent cations attract the oxygen ligands. The gain in Coulomb energy increases with increasing off-centre displacement. Table 4.10 proves that this expectation is correct. The largest effects are observed for silicon followed by aluminium.

Table 4.10. Off-centre displacement of extrinsic cations substituting for potassium in $KTaO_3$. The direction of the displacement corresponds to <100>. The energy gain is measured with respect to the on-centre position

Ion		Displacement (Å)	Energy gain (eV)
Li^+	a	0.64	0.04
Mg^{2+}	b	0.63	0.24
Fe^{2+}	b	0.44	0.05
Al^{3+}	b	1.15	2.62
Al^{3+}	c	1.10	1.72
Sc^{3+}	b	0.44	0.08
Ti^{3+}	b	0.68	0.56
V^{3+}	b	0.63	0.45
Cr^{3+}	b	0.85	1.09
Mn^{3+}	b	0.85	0.87
Fe^{3+}	b	0.87	0.86
Si^{4+}	b	1.48	9.18
Mn^{4+}	b	0.36	0.53
Ti^{4+}	c	0.37	0.69

[a] $Li^+ ... O^{2-}$ short-range potential from [247]
[b] Empirical potential parameters [158, 248]
[c] Electron gas potential parameters [248]

Of particular interest is the behaviour of trivalent iron substituting for potassium. Table 4.10 yields an off-centre displacement corresponding to 0.87 Å. From electron spin resonance two Fe^{3+}-spectra are known which can be related to the K site [234], i.e. one cubic and another strongly axial spectrum. The problem to be solved now is to explain the cubic ESR spectrum on the basis of an isolated off-centre Fe_K^{3+} cation. The superposition model may be employed to find "cubic solutions", which are compatible with an off-centre iron defect. In this context the term "cubic solutions" means that the axial parameter b_2^0 is close to zero. Such solutions can result from suitable combinations of O^{2-} and Fe^{3+} displacements. The results, which employ the Lennard–Jones-type radial function "A" in Fig. 4.2, are displayed in Fig. 4.3. These show that even pronounced off-centre displacements need not result in a large b_2^0 parameter due to cancellation of terms within the SPM. Analogous results are obtained with different radial function parametrizations. This is due to the observations that for all reasonable ion displacements in Fig. 4.3 the iron–oxygen separations are greater than 2 Å and that the various radial functions are similar in this regime. Thus, it may be that the "low coordination" minimum is responsible for the "cubic" spectrum, but further work will be needed to confirm this model. In particular it must be shown that the fourth-order axial ESR parameter b_4^0, which has not been considered so far, vanishes as well within a range of suitable ion relaxations.

It is finally noted that the observed axial ESR spectrum corresponding to $b_2^0 \approx 4.5\ \mathrm{cm}^{-1}$ cannot be assigned to the present off-centre iron defect. It has been argued [246] that, first, the necessary off-centre displacement should have been even larger than 0.87 Å, and, second and more importantly, one would have to expect unphysically large ligand relaxations (of the order of

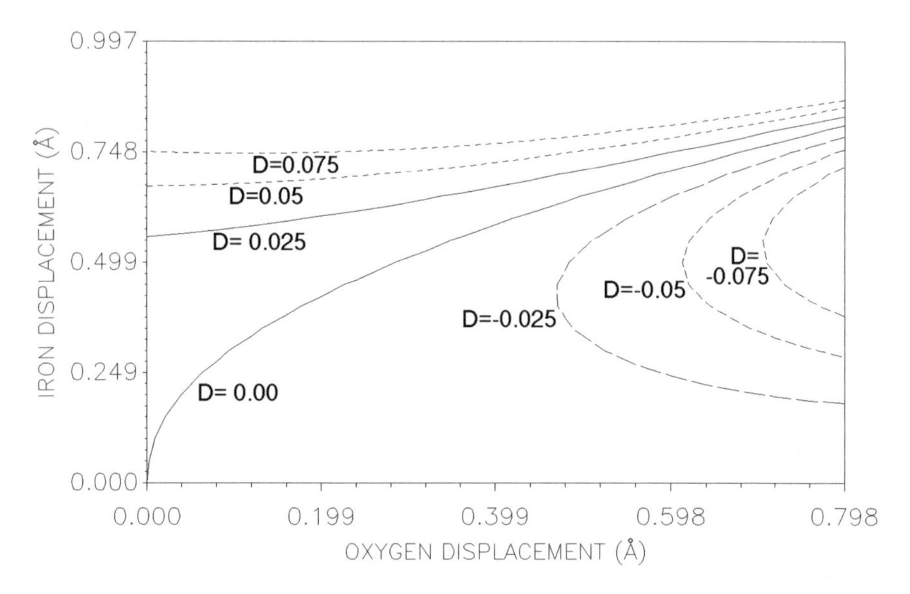

Fig. 4.3. ESR parameter b_2^0 relevant for Fe_K^{3+} obtained from a superposition model analysis. b_2^0 values (cm^{-1}) close to zero are displayed as functions of the iron off-centre and oxygen ligand displacement

1 Å). Such relaxations are not compatible with basic ion-size arguments and are not supported by shell model simulations. As is discussed in Sect. 4.2.2 it remains most reasonable to assign the axial ESR spectrum to the $Fe_K^{3+} - O_I^{2-}$ defect complex.

Nb_{Ta}^{5+} Off-Centre Defects in $KTaO_3$. In this subsection we consider possible off-centre displacements of Nb^{5+} ions which are incorporated in $KTaO_3$ at Ta sites. EXAFS measurements [249] performed on $KTa_{1-x}Nb_xO_3$ with $x = 0.09$ suggested a niobium off-centre displacement of ≈0.15 Å. Channelling experiments in KTN ($x = 0.03$) [250] could not detect any off-centre displacements within a reported accuracy of 0.1 Å. It is therefore tempting to provide predictions on the basis of atomistic simulations.

Recent theoretical investigations of Nb_{Ta} employed FLMTO-LDA supercell calculations [55]. These calculations have been performed at theoretical equilibrium structures which slightly underestimate the observed lattice spacings. In the case of KTN ($KTa_{1-x}Nb_xO_3$) the preferred [111] off-centre displacement of Nb has been found to be closely related to the TO phonon mode softening. However, the predicted critical niobium concentration $x \approx 0.22$ which leads to appreciable off-centre displacements referring to FE-PT is much too large compared with the experimental values of $x \approx 0.01 - 0.05$.

Shell model-based simulations, on the other hand, employing unconstrained lattice polarizations suggest that even isolated niobium impurities would be unstable with respect to off-centre displacements. The simulations are based on the shell model parametrizations discussed in Sect. 4.1. Figure 4.4 shows the Nb incorporation energy (renormalized to the on-centre position) as a function of an off-centre displacement along a [111] cubic crystal direction.

By inspection of the rigid-ion curves we infer that the *ab initio* Nb...O potential is *sui generis* better suited to model ionic off-centre displacements than the empirical interaction. However, both rigid ion calculations produce no appreciable off-centre displacements. Only if the complete shell model parametrization is employed the simulations indicate measurable niobium displacements, which is accomplished essentially by the niobium polarizability; oxygen polarizabilities, on the other hand, turn out to be ineffective in this respect. These results are valid using both Nb...O short-range potentials. The off-centre displacement of Nb ions is most pronounced in the case of the *ab initio* interaction. Using k_{Nb}=2500 eVÅ^{-2} one finds [111] and [110] off-centre displacements of about 0.08 Å, corresponding to an energy gain of 0.017 eV. However, it seems beyond the framework of ordinary shell model simulations to establish a clear distinction in favour of the observed [111] displacements. The calculated off-centre displacement becomes further increased ($\leq$ 0.15 Å) upon inserting k_{Nb}=2100 eVÅ^{-2}. Though this value will overestimate the off-centre behaviour of isolated Nb cations, it could receive some

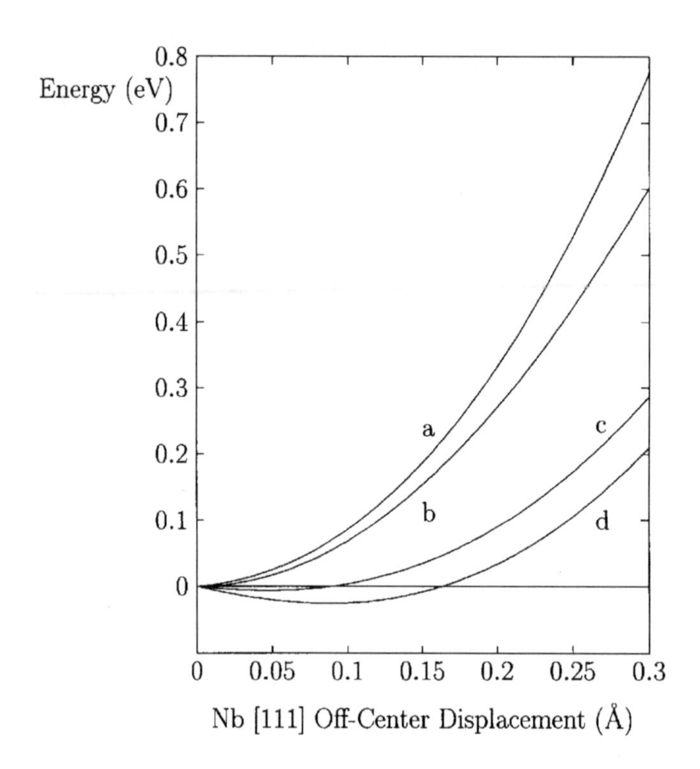

Fig. 4.4. Off-centre displacement (along a [111] cubic crystal direction) of one Nb^{5+} impurity incorporated in $KTaO_3$ on Ta site. All energies are renormalized with respect to the on-centre incorporation energy. (a) Empirical Nb^{5+}... O^{2-} potential with $k_{Nb}=\infty$. (b) *Ab initio* hydrostatic Nb^{5+}... O^{2-} potential with $k_{Nb}=\infty$. (c) Empirical Nb^{5+}... O^{2-} potential with k_{Nb}=1358.58 eV/Å^2. (d) *Ab initio* hydrostatic Nb^{5+}... O^{2-} potential with k_{Nb}=2500.0 eV/Å^2

relevance at higher doping levels[2]. In any case it clearly demonstrates the effectiveness of the niobium polarizability. Importantly, the present simulations also confirm that there are no off-centre displacements in the case of tantalum when using the shell model parametrization for $KTaO_3$ given in Sect. 4.1. All results may be easily interpreted due to important covalency effects. As has been argued in Sect. 4.1, covalency is expected to be stronger in the case of niobium ions than in that of tantalum ions, which may be expressed effectively by an enhanced niobium polarizability. It should be stressed that both the electronic polarizability of Nb ions as well as lattice relaxations are necessary for possible off-centre displacements to occur. Indeed, all model

[2] The idea of concentration-dependent spring constants is not unfamiliar; it has also been employed in earlier lattice-dynamical investigations on KTN [251]. However in that case the oxygen polarizability was varied upon changing the Nb content.

calculations in which exactly one of these features is taken into account fail to give a non-vanishing off-centre displacement. Further, it is not sufficient to include only the relaxations of neighbouring ions. Shell model simulations and preliminary embedded cluster DFT-BLYP calculations employing a relaxed $NbO_6K_8Nb_6$ cluster embedded in a perfectly structured lattice clearly favour the on-centre configuration. This result indicates that important elastic effects (lattice-deformation fields) facilitate the niobium displacement at low concentrations. Embedded cluster calculations are in progress in order to address these questions in further detail. However, one should note that the present calculation of possible off-centre displacements ignores any effects due to thermally induced vibrations and zero-point motions. Including such influences could reduce the actual niobium off-centre displacement. In conclusion, it seems reasonable that the combination of extended lattice deformations induced by the niobium off-centre defect and coupling to the soft TO phonon explains the smallness of the observed critical niobium concentrations in solid solutions KTN.

Strong support in favour of the present calculations is given by very recent inelastic neutron scattering data [252] in dilute KTN crystals ($x \approx 0.01$). The parabolic dispersion behaviour of the soft TO phonon measured for small wave vectors $\boldsymbol{q}$ along a cubic crystal axis, i.e.

$$\omega(\boldsymbol{q})^2 = \omega_0^2(1 + r_c^2\boldsymbol{q}^2)\,, \tag{4.17}$$

indicates a very pronounced increase of the correlation length r_c upon reducing the temperature. This behaviour is absent in pure $KTaO_3$. At $T \approx 10K$ r_c becomes five times as large as in pure $KTaO_3$ corresponding to about 80 Å. The data have been interpreted due to existing polar clusters around the niobium impurities in cubic $KTaO_3$. r_c measures the size of these polar clusters. The results agree with the assumption of off-centre niobium defects inducing extended lattice deformations. Note that the experiment cannot be explained on the basis of traditional shell model simulations, which employ a variation of the oxygen polarizability and the virtual crystal approximation to account for the niobium solution (see also [251]).

Li_K^+ Off-Centre Defects in $KTaO_3$. Doping of $KTaO_3$ with monovalent Li cations also causes the formation of polar phases. The critical concentration x_c in KLT ($K_{1-x}Li_xTaO_3$) has been reported to be 0.01 (e.g. [217]), thus being similar to KTN discussed in the preceding subsection. The comparable effectivity of Li doping refers to the pronounced off-centre displacement (along $\langle 100 \rangle$ directions) in this case (see below). On the other hand, the alternative incorporation of sodium, which only shows a very small off-centre displacement of 0.04 Å [217], leads to a significantly increased critical concentration of $x_c = 0.13$.

Shell model simulations [206, 253] and LDA supercell investigations [47, 55] consistently predict Li off-centre displacements along [100] ranging from 0.61–0.64 Å in [47, 55, 206] to 1.44 Å in [253]. Shell model approaches may suffer from significant dependences upon Li impurity potential parameters. For example, slightly inconsistent partial charge models for $KTaO_3$ and $LiKSO_4$ (from which the short-range Li...O defect potential has been taken) used in [253] lead to a pronounced overestimation of the off-centre displacement (see [206] for details). LDA results, on the other hand, are influenced by supercell sizes and by the neglect of lattice relaxations. Embedded cluster simulations are in progress to clarify the situation.

We now consider some details obtained from shell model simulations on isolated Li_K^+ defects. The necessary Li...O short-range impurity–oxygen interaction and Li shell parameters are taken from shell model investigations on Li_2O [247]. These parameters were fitted to adjust structural as well as lattice dynamical data and have been successfully used in ion transport studies [254]. This impurity parametrization is also assumed to reliably model the off-centre behaviour of lithium cations incorporated in $KTaO_3$.

Figure 4.5 confirms that the off-centre displacement occurs along the <100> direction. This is in agreement with results from nuclear magnetic resonance (NMR) experiments, according to which the lithium ion is shifted a distance ranging from 0.86 Å [255] to 1.1 ± 0.1 Å [220] along the $\langle 100 \rangle$ direction. The calculated Li^+ off-centre displacement of 0.64 Å is less than these values, but it should be considered that in the evaluation of the experiment assumptions concerning the environment of the defect are implicit. The isotope ^7Li with a nuclear spin of $\frac{3}{2}$ posesses a quadrupole moment which couples to the electric field gradients. Therefore an NMR measurement of the level splitting provides information on the local electric field gradients acting on the ion observed. On the basis of certain assumptions, e.g. formal ion charges and the perfect lattice positions of ions, the off-centre position of the Li^+ ion can be inferred from the splitting. Borsa et al. [255] report an NMR frequency for the ^7Li off-centre ion of 70±2 kHz, from which they deduced their value for the off-centre shift of 0.86 Å. By calculating the electric field gradient V_{zz} employing the present relaxed environment, which is chosen to include all displaced ions within a radius of 3.3 lattice units of the Li off-centre defect, the NMR frequency may be obtained using the formula:

$$\nu_Q = |(1 - \gamma_\infty) e Q V_{zz}| \, . \tag{4.18}$$

The quadrupole moment of ^7Li is given by $Q = 0.042\ 10^{-24}$ cm^2 and $(1 - \gamma_\infty) = 0.74$ determines the asymptotic Sternheimer shielding factor [256]. The experimental nuclear magnetic resonance frequency corresponds to an Li position of 0.78 Å. If the calculation is performed for the unrelaxed lattice one obtains a position of 1.1 Å corresponding to 70 kHz. Thus, relaxation effectively results in a reduction of the calculated Li^+ off-centre displacement. Conversely, for the present equilibrium position of ca. 0.64 Å a frequency of 25 kHz is calculated. The strong variation of frequency shows the sensitivity of the NMR frequency on the Li off-centre position, which is in agreement with calculations of van der Klink et al. [256].

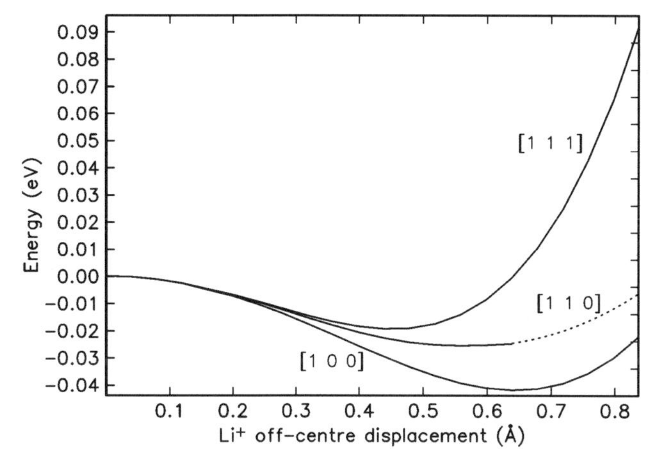

Fig. 4.5. Defect energy versus off-centre displacement for Li^+ in $KTaO_3$ in different directions. The dashed part of the line corresponds to calculations which would not converge – possibly due to the vicinity of the lithium ion to oxygen ions in this direction. The curve shown is an extrapolation of a polynomial fit to the converged part of the curve. All energies are renormalized with respect to the on-centre position

Figure 4.5 also shows that $\pi/2$ jumps crossing the $\langle 110 \rangle$ direction are the most likely way of reorientating the defect. The energy for the $\pi/2$ reorientation is ca. 0.015 eV (120 cm^{-1}), as can be seen by taking the difference of the energy gain on relaxing in the $\langle 100 \rangle$ and the $\langle 110 \rangle$ direction. Though this value is also found from SHG (second harmonic generation) measurements on 0.8% Li-doped $KTaO_3$ crystals [257], it still remains open of whether the experimental value corresponds to the simple hopping process discussed in the present shell model simulations. Indeed, it has recently been argued that the observed activation energy might be related to resonance tunnelling via localized excited vibrational states, which become suppressed with increasing Li concentrations [258]. Anticipating the validity of this resonant tunelling interpretation would mean that the present shell model simulations slightly underestimate the hopping barrier. However, in any case the shell model prediction seems to indicate the correct order of magnitude, i.e. some tens meV.

In Fig. 4.6 the energy surface of the Li_K^+ off-centre defect is shown versus the displacement in the (100) plane. Clearly the central on-centre energy maximum and the four off-centre minima along the $\langle 100 \rangle$ directions in the plane can be seen, as well as the $\langle 110 \rangle$ saddlepoint directions.

The local defect geometry qualitatively resembles the picture shown in Stachiotti et al. [253]. The displacements indicated there, though, are considerably larger than the ones calculated in this work. The discrepancy might be due to the harmonic approximation applied in [253]. The general pattern is such that anions are attracted towards and cations are repelled from the positive (Li^+) end of the defect dipole. The opposite movements are seen

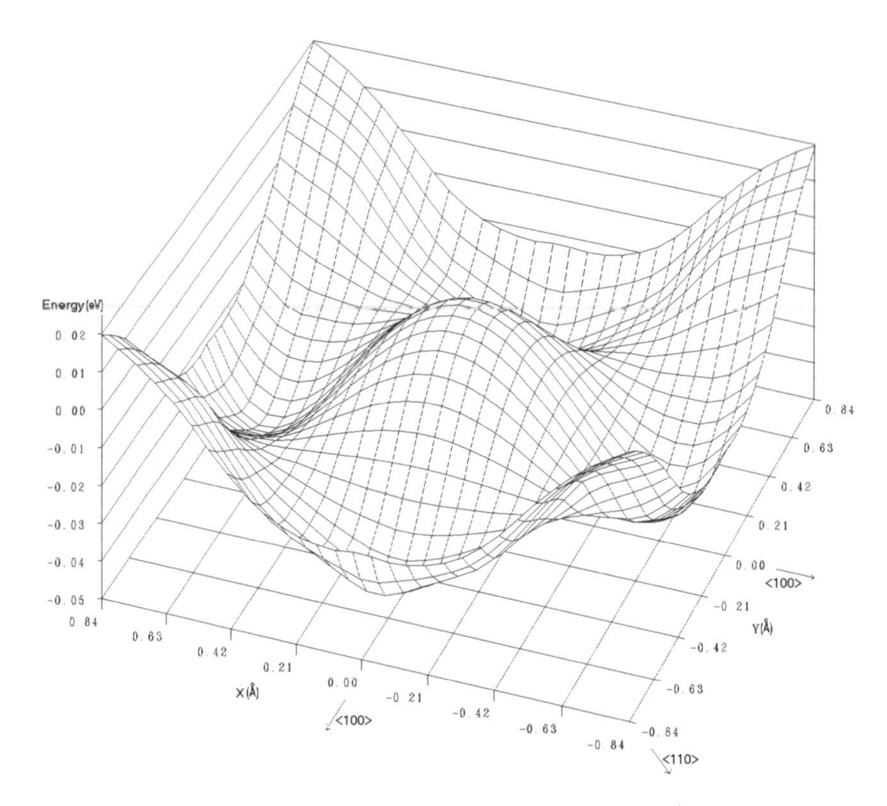

Fig. 4.6. Defect energy versus off-centre displacement for Li_K^+ in $KTaO_3$ over the (100) plane. Four off-centre minima along ⟨100⟩ and saddlepoint directions along ⟨110⟩ are shown as well as the central on-centre maximum

at the negative (V_K) end of the dipole. It is noted, in addition, that Li_K^+ off-centre defects represent effective hole traps in $KTaO_3$. The present shell model simulations indicate a hole (O_O^-) binding energy of 0.3 eV, of which 0.13 eV is due to the ion-size misfit of lithium and 0.17 eV is due to the off-centre displacement.

The off-centre behaviour of the lithium ion can be considered under various conditions, i.e.

- without allowing for lattice relaxation,
- allowing for lattice relaxation, but treating the the $KTaO_3$ host crystal within the rigid ion framework,
- treating Li^+ as an unpolarizable (rigid) ion within a shell model host.

If lattice relaxation is not admitted (neither cores nor shells relax) no lithium off-centre displacement occurs. Therefore lattice relaxation seems to be indispensable for any off-centre displacement. But importantly, there is no measurable off-centre displacement if the host is treated within the rigid ion model, even if lattice relaxations are included. Moreover, if only the lithium ion is unpolarizable, off-centre displacements are still present, but they are far smaller in extent and energy gain. These simulations indicate that the electronic structure plays an important role in facilitating the occurrence of lithium off-centre displacments. This conclusion is also supported by preliminary embedded cluster calculations. Figure 4.7 displays the calculated off-centre behaviour using embedded cluster DFT-BLYP calculations (see also Sect. 2.1) with CADPAC [173]. In these preliminary investigations all ions except the Li cation are held fixed at their perfect lattice positions. Split valence Gaussian-type orbitals are employed for lithium and its 12 oxygen ligands and bare effective core potentials to represent the next-nearest cation neighbours. In addition, polarizing d-functions are added to the oxygen basis set. It is emphasized that the observed off-centre displacement does not occur if the Hartree–Fock approximation is employed. This result supports the importance of electronic structure effects, which are also evident from the shell model calculations discussed above. But similarly to the HF simulations, the shell model is unable to produce any Li off-centre displacements if only the shells of ions are allowed to relax (so-called "optical" simulations). In comparison with shell model simulations the preliminary embedded cluster calculations slightly underestimate the Li off-centre displacement. This discrepancy can probably be remedied upon further improving the basis set quality. Very recent FLMTO supercell simulations yield Li off-centre displacements of 0.64 Å [259] which are in good agreement with previous shell model simulations of Exner et al. [206]. The preliminary inclusion of lattice relaxations in the FLMTO set of simulations suggests a delicate dependence of the off-centre energy gain upon relaxations, whereas the off-centre displacement seems to remain unaffected.

Investigations of Li–Li interactions [207, 260] indicated that antiferroelectric longitudinal dipole ordering is most favourable at short Li–Li separations which emphasizes the importance of elastic interactions mediated by the host lattice. Electrostatically one would expect a ferroelectric alignment of the Li off-centre dipoles due to a classical dipole–dipole interaction. Upon increasing the Li–Li separation a ferroelectric transverse dipole ordering becomes preferred. Stachiotti and Migoni [260] investigated dependences upon the Li concentration employing their earlier shell model parametrization [253]. Although the Li...O short-range potential used can be criticized (see above), the investigations seem to provide useful trends concerning the interaction of different Li cations. Effects depending on the distance, the bond direction and on the dipole orientation have been studied on the basis of the effective Hamiltonian

$$H = -\frac{1}{2}\sum_{i,j}\sum_{\mu,\nu} J_{i,j}^{\mu,\nu} n_i^{\mu} n_j^{\nu} , \qquad (4.19)$$

which is confined to include two-body interactions. Sites are denoted by i and j, whereas greek indices characterize the six dipole orientations; n_i^{μ} is

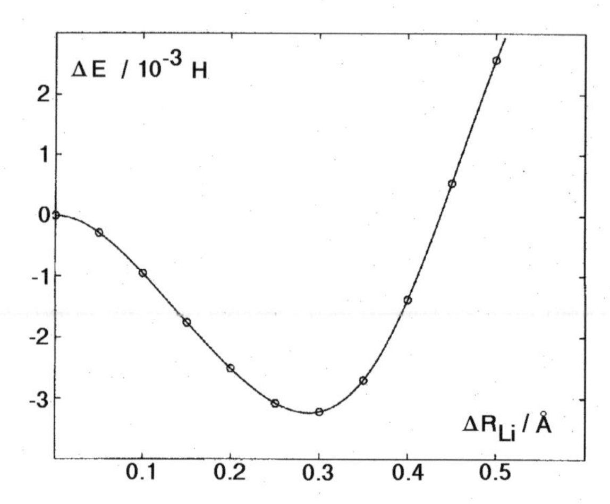

Fig. 4.7. The defect energy derived from perfect lattice embedded cluster calculations is shown as a function of the Li off-centre displacement along [100]. The calculations employ density functional theory using the GGA-type BLYP exchange correlation functional. The energies are renormalized with respect to the on-centre position

a random occupation variable. The calculations suggest a tendency towards dipole ordering at high concentrations. However, as long as two-body interactions are assumed to be dominant, frustration effects remain large even at high concentrations (see also [219] for related information).

In this context, we also note recent preliminary semi-empirical INDO supercell simulations of $KTaO_3$:Li [261], which were used to probe the Li–Li interaction in this material. The simulations of isolated Li off-centre defects support the formation of a lattice polarization cloud, which enhances the effective dipole moment of the off-centre defect. Calculations of interacting Li cations indicate that the Li–Li interaction strength decreases significantly faster with increasing Li–Li distance than predicted by the earlier shell model simulations of Stachiotti et al. [260]. Since all simulations of interacting Li cations neglect lattice polarization contributions (i.e. only electronic polarizations are accounted for by means of INDO), the interaction pattern fits the classical dipole–dipole interaction. However, the inclusion of lattice polarizations probably modifies the interaction behaviour.

4.2.2 Dipole Centres Due to Defect Aggregation

Defect Complexes Involving Oxygen Vacancies. This subsection is devoted to the structure of defect complexes involving oxygen vacancies trapped at acceptor impurity ions on the Ta or Nb sites of $KTaO_3$ and $KNbO_3$, respectively. However, the calculated results are of general significance and may be safely extended to other defect complexes of this type. It is noticed in this context that $M_B^{n+} - V_O$ dipoles are frequently observed in all ABO_3 perovskites. Their formation may be explained by aggregations between impurity cations and oxygen vacancies which are created as intrinsic charge compensators or are due to some reducing crystal growth conditions. Alternatively, one may discuss the impurity-assisted Frenkel defect formation (see Sect. 4.1.2).

Prominent examples of these defect complexes are Fe^{3+}–V_O and $Mn^{2+} - V_O$. Both paramagnetic defect centres are characterized by an axial ESR parameter b_2^0 around 1 cm^{-1}. This value may be explained through the cation displacement relative to the oxygen vacancy. Earlier investigations employing the superposition model suggested that the metal cation moves towards the vacancy [239, 262] (which would significantly reduce its dipole character). At the end of this subsection we shall discuss a refined SPM analysis which, in agreement with shell model and embedded cluster calculations, favours the opposite iron displacement. This result emphasizes the particular role of such complexes as effective polarizing lattice perturbations.

Shell model-based results for $KTaO_3$ are shown in Table 4.11. All numbers in this table are obtained using empirical impurity–oxygen potential parameters [158, 248].

It becomes clear that almost all cations move away from the oxygen vacancy. The only exceptions to this general rule are provided by the very large monovalent cations Na^+ and Rb^+. In these latter two cases the electrostatic attraction between the cation and its fivefold ligand sphere is not sufficiently large to overcompensate for the corresponding short-range repulsion between these ions. From Table 4.11 it emerges as a general trend that the binding energy increases with decreasing charge of the cation.

Next we consider in some more detail the $Fe - V_O$ complex and in particular the axial displacement of the iron cation as a function of its *formal* charge state. The short-range Fe...O interaction is fixed and taken as the empirical Fe^{3+}... O^{2-} potential derived by Lewis and Catlow [158]. This procedure, of course, ignores variation of the short-range potential with the charge of the cation, but these effects are unlikely to have a large effect on the present results. The calculations demonstrate the pronounced tendency for the iron cation to be displaced away from the oxygen vacancy. Figure 4.8 shows the calculated displacements of the Fe^{n+} cation along the axial direction in $KTaO_3$ and $KNbO_3$. It is clear that the iron cation moves in the opposite direction of the oxygen vacancy for charge states between +1 and +5. Only with unusual charge states outside this range does the iron cation tend to move towards the vacancy, especially in $KTaO_3$; such charge states will of course not be observed in any real system. However, results for these hypothetical states do give us insight into the factors controlling the direction of the displacements.

For low charge states the electrostatic attraction between the iron cation and the remaining fivefold oxygen polyhedron is small. Therefore, in order

Table 4.11. Binding of oxygen vacancies V_O to extrinsic metal cations M on tantalum site in $KTaO_3$. The displacement in the table refers to the axial relaxation of the metal ion. Negative energies correspond to a bonding situation between both defects; positive relaxations denote a displacement of the cations away from the vacancy

Ion M	Binding energy (eV) $M_{Ta} - V_O$	Displacement (Å)	Ion M	Binding Energy (eV) $M_{Ta} - V_O$	Displacement (Å)
Li^+	−1.19	0.25	Mn^{3+}	−0.53	0.34
Na^+	−1.36	−0.05	Fe^{3+}	−0.57	0.35
Rb^+	−2.83	−0.93	Y^{3+}	−0.16	0.22
Mg^{2+}	−0.75	0.31	La^{3+}	−0.32	0.15
Fe^{2+}	−0.72	0.29	Nd^{3+}	−0.23	0.18
Mn^{2+}	−0.72	0.26	Eu^{3+}	−0.19	0.20
Ca^{2+}	−0.77	0.16	Gd^{3+}	−0.18	0.20
Sr^{2+}	−1.04	0.08	Ho^{3+}	−0.16	0.22
Al^{3+}	−0.84	0.37	Yb^{3+}	−0.17	0.24
Sc^{3+}	−0.29	0.30	Lu^{3+}	−0.17	0.24
Ti^{3+}	−0.35	0.31	Si^{4+}	0.72	0.38
V^{3+}	−0.31	0.30	Mn^{4+}	−0.04	0.29
Cr^{3+}	−0.48	0.33	Zr^{4+}	0.42	0.24

to reduce the short-range repulsion the iron cation is displaced towards the vacant oxygen site. Essentially, this is a space-filling effect and also applies to Na^+ and Rb^+ discussed above. The small iron charge in this case allows the "occupation" of an anionic lattice site. The situation changes drastically for higher charge states. On the other hand, for extremely large charge states the electrostatic attraction is so pronounced that the whole ligand sphere is considerably contracted, which results in the observed iron displacement towards the vacancy.

Subsequently, we consider further details in the case of trivalent iron. Shell model and embedded cluster calculations exemplify the situation for $Fe^{3+}_{Nb} - V_O$ dipole centres in $KNbO_3$. Figure 4.9 visualizes the defect-induced ion relaxations. The pattern is obtained from shell model simulations, but note that embedded cluster calculations (see below) yield almost identical results. In particular, the separation between iron and the axial oxygen ligand opposite the vacant oxygen site becomes effectively reduced from 2 Å in the perfect lattice to 1.82 Å upon relaxations. One should define an effective off-centre displacement $\delta^{\text{eff}}_{\text{Fe}}$ which relates the iron displacement along the defect z-axis to the z-displacement of the four planar oxygen anions, i.e.

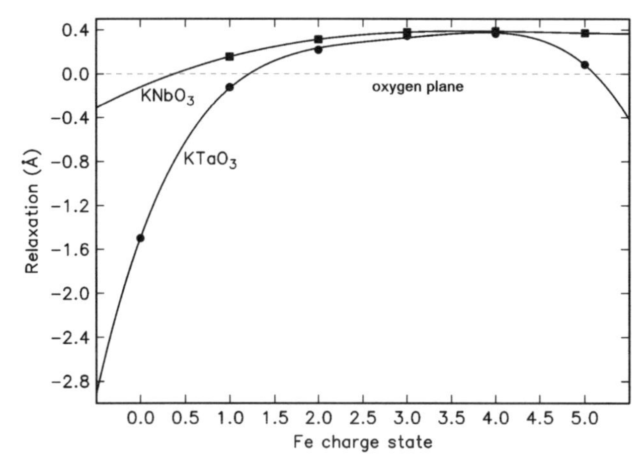

Fig. 4.8. Relaxation of an Fe cation relative to an associated oxygen vacancy as a function of the Fe charge state and with the short-range cation–oxygen interaction being fixed to the empirical Fe^{3+}... O^{2-} potential. The results refer to $KTaO_3$ and $KNbO_3$ as host crystals. The dashed line denotes the position of the octahedral oxygen plane containing four equatorial ligand anions; this plane is perpendicularly oriented to the main Fe^{3+}–V_O axis. Positive relaxations denote a displacement of the iron cation away from the vacancy

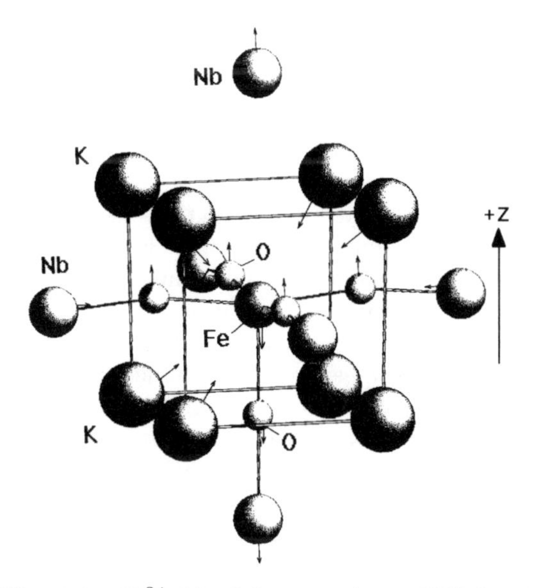

Fig. 4.9. Fe^{3+}–V_O defect complex in $KNbO_3$ with its nearest oxygen, niobium and potassium neighbours. The relaxed positions have been calculated on the basis of a shell model simulation. Arrows denote the directions of ionic displacements

$$\delta_{\mathrm{Fe}}^{\mathrm{eff}} = \delta z(\mathrm{Fe}) - \delta z(\mathrm{O}_{xy}) \,. \tag{4.20}$$

This renormalization is reasonable due to the possibly pronounced oxygen relaxations (see Fig. 4.9) In this context we should remember that SPM-based analyses can only yield the relative displacements between the iron and its relaxed oxygen ligands due to a locally confined scan of the environment geometry. Thus one should compare these relative SPM and shell model displacements representing the relevant entities. Shell model calculations yield $\delta_{\mathrm{Fe}}^{\mathrm{eff}} \approx -0.4$ Å with negative values indicating a displacement away from the vacant oxygen site. This value is obtained with both Nb...O potential options.

The general lattice relaxation pattern emerging from these simulations may be interpreted following basic screening arguments: electronic as well as lattice relaxations tend to screen the two charged ends of the dipole defect complex considered. To put it simply, the vacancy attracts anions and repells cations; the Fe^{3+}, on the other hand, repells anions (compared with the substituted Nb^{5+}) but tries to retain a pure anionic coordination sphere. In view of these relaxations it is again obvious that the iron cation avoids occupying the vacant anionic lattice site, as this would considerably increase the *local* excessive defect charge resulting in costly elastic energy terms in order to screen the resulting defect charge.

Sophisticated simulations may be performed using the embedded cluster technique. The quantum defect cluster actually chosen corresponds to the ion configuration shown in Fig. 4.9. It consists of a $(\mathrm{FeO}_5)^{7-}$ fragment and of eight potassium and six niobium ions at the cluster boundary. MO descriptions are employed for the ions belonging to the FeO_5 complex with Gaussian basis functions for iron [125] and for the oxygen ions [126] corresponding to split valence (SV) quality. In addition, the oxygen basis set is augmented with polarizing d-functions. Moreover, s- and p-type functions (with exponents 0.01, 0.1 and 1.0 for both types) are implemented at the vacant oxygen site. Finally, the cations at the cluster boundary are represented by effective core pseudopotentials [21]. The *ab initio* cluster calculations are performed at the UHF and DFT levels using the quantum chemical programs HONDO [172] and CADPAC [173]. The exact exchange correlation functional is approximated with the GGA-type BLYP functional (see Sect. 2.1).

The embedding lattice is modelled using the shell model parametrization referring to the *ab initio* Nb...O short-range potential (see Sect. 4.1). As was discussed in [263], the defect-induced ion relaxations do not depend sensitively on the particular choice of k_{Nb}, but $k_{\mathrm{Nb}} = 2500$ eVÅ^{-2} is certainly a good choice. Pair potentials are also employed to simulate the cluster–lattice short-range interactions. The implementation of geometry optimizations follows the general procedure discussed in Sect. 2.1.2. Formal charges are used in the point charge cluster representation (see Sect. 2.1.2), leading to an approximate dipole consistency. Due to the exact cluster relaxation the present calculations improve on earlier results reported in [263], but the deviations remain small.

In a first set of calculations the relaxation of the Fe^{3+} cation is investigated assuming an otherwise unrelaxed (i.e. perfect) lattice (Fig. 4.10). As was discussed above the crystal tends to screen the excessive defect charges of the iron ion and of the oxygen vacancy.

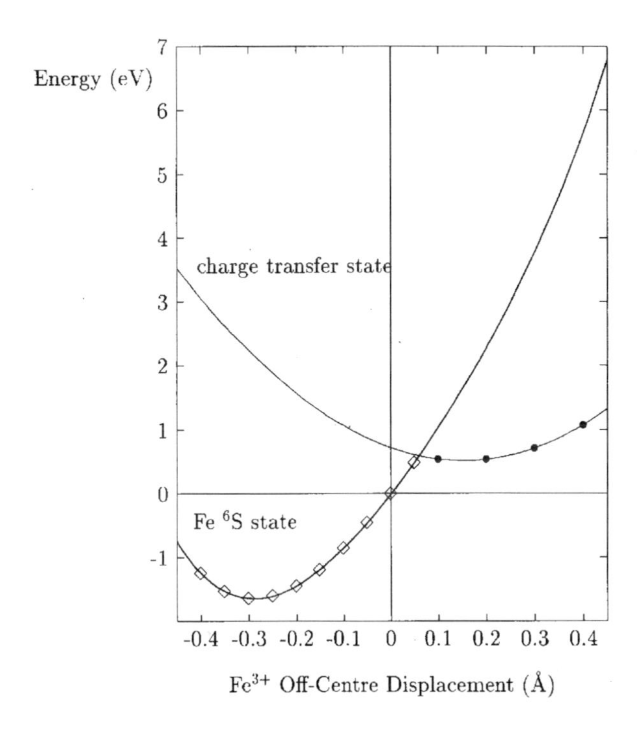

Fig. 4.10. Embedded cluster simulation of Fe^{3+}–V_O within an otherwise perfect and unrelaxed lattice. Only the iron cation is allowed to be displaced

Within a perfect lattice, in which only the iron cation is allowed to be displaced, there are two choices of defect screening: the displacement of the iron ion along $-z$ and an electron transfer onto the oxygen vacancy. It is noted that the latter screening mechanism cannot be accounted for in shell model simulations. Displacing the iron towards the centre of the remaining fivefold oxygen polyhedron effectively optimizes the anionic environment of this cation in comparison with the opposite relaxation towards the vacancy. Figure 4.10 confirms that a shift of the iron ion towards the associated vacancy is energetically highly unfavourable. Indeed, this would considerably increase the excessive positive defect charge at the vacant oxygen site. Most favourable is the iron displacement $\delta_{\mathrm{Fe}}^{\mathrm{eff}} \approx -0.3$ Å, which is in good agreement with the shell model calculations.

The alternative screening mechanism involves an electron transfer from the oxygen ligands of iron onto the vacancy while maintaining the iron ^{6}S state (a change of the high-spin configuration of Fe^{3+} turns out to be highly unfavourable). By inspection of Fig. 4.10 we observe that the electron transfer is considerably less favourable than an alternative displacement of the iron along $-z$. The energy dependence of the charge transfer state has a minimium for iron displacements $\delta_{\text{Fe}}^{\text{eff}} \approx +0.15$ Å indicating that in this situation the fivefold oxygen coordination sphere becomes less attractive for iron.

The reason that electron transfer represents an energetically costly screening mechanism is the high position of the vacancy-related electron states in comparison to the Fe_{Nb}^{3+} impurity states. Lattice relaxations, which are considered next, further destabilize electrons localized in genuine vacancy orbitals.

Table 4.12 compiles the calculated ion relaxations within the FeO_5 fragment of the total quantum cluster. By inspection one finds $\delta_{\text{Fe}}^{\text{eff}}(\text{UHF}) = -0.35$ Å and $\delta_{\text{Fe}}^{\text{eff}}(\text{DFT} - \text{BLYP}) = -0.29$ Å. Both values agree favourably with the perfect lattice calculations but also with results from shell model simulations.

Mulliken population analyses (MPA) confirm that lattice relaxations destabilize the electronic charge density at the vacant oxygen site. For example, on employing a perfect lattice geometry and $\delta_{\text{Fe}}=0$ the negative electronic charge of the oxygen vacancy corresponds to $-0.4\ |e|$ within UHF. If we include completely relaxed ion positions the electronic vacancy charge is reduced to about $-0.05\ |e|$. Similar results are obtained with DFT, but the major effects are due to an increase of covalent charge transfer within the FeO_5 fragment.

Table 4.12. Calculated ion relaxations in the FeO_5 complex. O_{xy} denotes one representative of the four planar oxygen anions and O_z the axial ligand ion. The displayed coordinates refer to cubic crystal axes

Ion	x/Å	y/Å	z/Å
	UHF		
Fe	0.0	0.0	−0.23
O_{xy}	1.96	0.0	+0.12
O_z	0.0	0.0	−2.05
	DFT-BLYP		
Fe	0.0	0.0	−0.18
O_{xy}	1.98	0.0	+0.11
O_z	0.0	0.0	−2.03

Analogously, lattice relaxation also causes a pronounced energy shift of the charge transfer state with respect to the ^{6}S ground state: According to previous calculations [263] the charge transfer state is shifted 3 eV above the ground state at $\delta_{\mathrm{Fe}}^{\mathrm{eff}} \approx 0.0$ Å. Approximately this situation is accomplished through a parallel displacement of iron and the planar oxygen ions by 0.3-0.4 Å towards the vacancy[3]. The corresponding perfect lattice energy difference at $\delta_{\mathrm{Fe}}^{\mathrm{eff}} = 0$ is about 0.8 eV (see Fig. 4.10).

In summary, shell model as well as embedded cluster calculations unambiguously prove that Fe^{3+} cations substituting for niobium in $KNbO_3$ do not move towards an associated oxygen vacancy. The calculations also indicate that this result can be generalized to all defect complexes of this type. In passing we quote additional results on $Ti_{\mathrm{Ti}}^{3+} - V_{\mathrm{O}}$ (i.e. the F^+ centre, see also Sect. 3.3.4) and $Mn_{\mathrm{Ti}}^{2+} - V_{\mathrm{O}}$ in $BaTiO_3$. The corresponding effective cation displacements are -0.36 Å for titanium[4] upon complete lattice relaxation and -0.25 Å for managnese, assuming an otherwise unrelaxed lattice. Consequently, earlier SPM interpretations must be modified in order to achieve consistency with these independent atomistic calculations.

Thus we now consider a refined SPM analysis. Although formulated for the special situation in $KNbO_3$, it is indeed general enough to be applied to all S state impurities which are associated with oxygen vacancies in ABO_3 perovskites. From (4.14) one obtains the following expression describing the axial ESR parameter b_2^0 of $Fe^{3+} - V_{\mathrm{O}}$ as a function of the effective iron displacement measured with respect to the oxygen ligands:

$$b_2^0(\delta_{Fe}^{\mathrm{eff}}) = 2\bar{b}_2\left(\sqrt{a^2 + (\delta_{Fe}^{\mathrm{eff}})^2}\right)\left[3\left(\frac{\delta_{Fe}^{\mathrm{eff}}}{\sqrt{a^2 + (\delta_{Fe}^{\mathrm{eff}})^2}}\right)^2 - 1\right] + \bar{b}_2(a + \delta_{Fe}^{\mathrm{eff}})\,. \tag{4.21}$$

$\delta_{\mathrm{Fe}}^{\mathrm{eff}}$ denotes the effective off-centre displacement of the iron cation defined in (4.20) and "a" the perfect iron–oxygen separation. The ligand configuration corresponds to the perfect lattice positions in cubic $KNbO_3$. This assumption does not substantially restrict the results, but simplifies the analysis. The first term on the right-hand side in (4.21) represents the contribution of the four planar oxygen ligands, whereas the last term stems from the axial oxygen ion. The first term dominates for large positive off-centre displacements (i.e. towards the vacant site); the second one, in turn, dominates for increasingly negative displacements.

The present SPM analysis includes an inverse power law and a Lennard–Jones-type radial function, i.e.

[3] The energy separation certainly becomes larger for $\delta_{\mathrm{Fe}}^{\mathrm{eff}} < 0$; however, the charge transfer state could not be stabilized in such cases.

[4] This value (-0.36 Å $= -0.21$ Å (Ti) $-$ 0.15 Å (O_{xy})) refers to titanium tacitly assumed in a t_{2g} state. However, the corresponding effective displacement becomes -0.14 Å if titanium is taken in its delocalized e_g-type ground state.

$$\bar{b}_2(r) = \bar{b}_2(R_0)\left(\frac{R_0}{r}\right)^{t_2}$$

and

$$\bar{b}_2(r) = A\left(\frac{R_0}{r}\right)^{M} + B\left(\frac{R_0}{r}\right)^{N},$$

respectively.

Due to charge misfit effects (note that Fe^{3+} substitutes for Nb^{5+}) $\bar{b}_2(R_0)$ becomes more negative by 22% of its original value, i.e. the actual constant is set equal to -0.5 cm^{-1}. These changes are in line with investigations of Müller and Berlinger [243] which were based on earlier shell model-type calculations of Sangster [177]. The Lennard–Jones-type radial function (with A=0.8 cm^{-1}, B=0.3 cm^{-1}, N=11 and M=16) is chosen to fit with the inverse power function at 2 Å, but shows significant deviations at smaller iron–ligand separations. Figure 4.11 compares the two actual $\bar{b}_2(r)$ dependences.

Figure 4.12 displays the resulting b_2^0 parameters as a function of the effective iron displacement. Whereas the inverse power law fits with the experimental value only for $\delta_{\text{Fe}}^{\text{eff}} \approx +0.25$ Å, the Lennard–Jones-type dependence

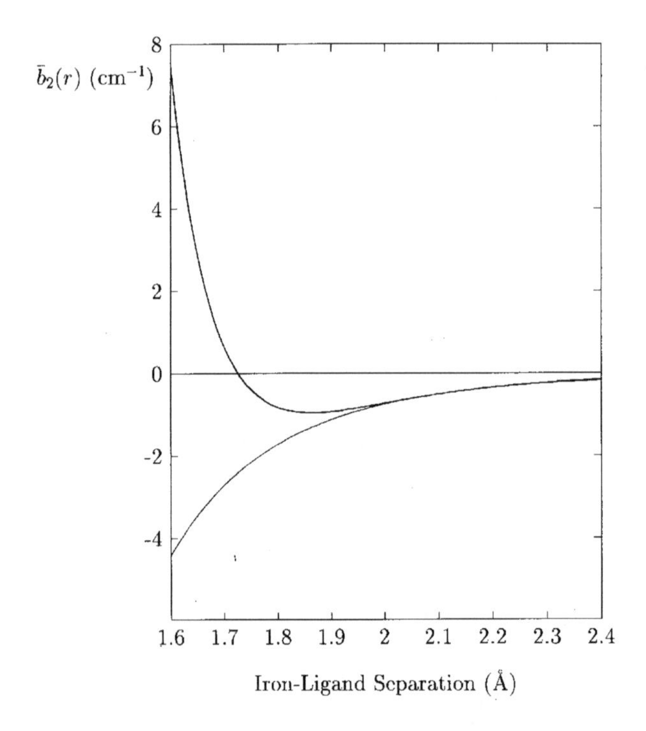

Fig. 4.11. Comparison of the inverse power function with a Lennard–Jones-type dependence

allows for a second match with experiment at $\delta_{\mathrm{Fe}}^{\mathrm{eff}} \approx -0.25$ Å. Only this second solution referring to the Lennard–Jones-type function is qualitatively consistent with shell model and embedded cluster calculations.

$Fe_K^{3+} - O_I^{II}$ Complexes in $KTaO_3$. ESR experiments have shown that the Fe^{3+} ion can reside on both cation sites in $KTaO_3$. In addition to cubic-type spectra [234], which are assigned to the dopant ion substituting for both cations with remote charge compensation, axial-type spectra are observed. These usually are ascribed to $Fe_{Ta}^{3+} - V_O$ and $Fe_K^{3+} - O_I$, V_O being an oxygen vacancy and O_I an oxygen ion on the nearest interstitial site to the Fe_K^{3+} [264]. The subsequent analysis, combining shell model and superposition model investigations, strongly supports this interpretation of the K site axial centre. Very recent experiments [222] on these centres based on optically detected magentic resonance techniques agree with this assignment too.

The ground state axial zero-field-splitting parameter b_2^0 is measured as 1.33 cm^{-1} for the Ta site axial centre and 4.46 cm^{-1} for the K site axial

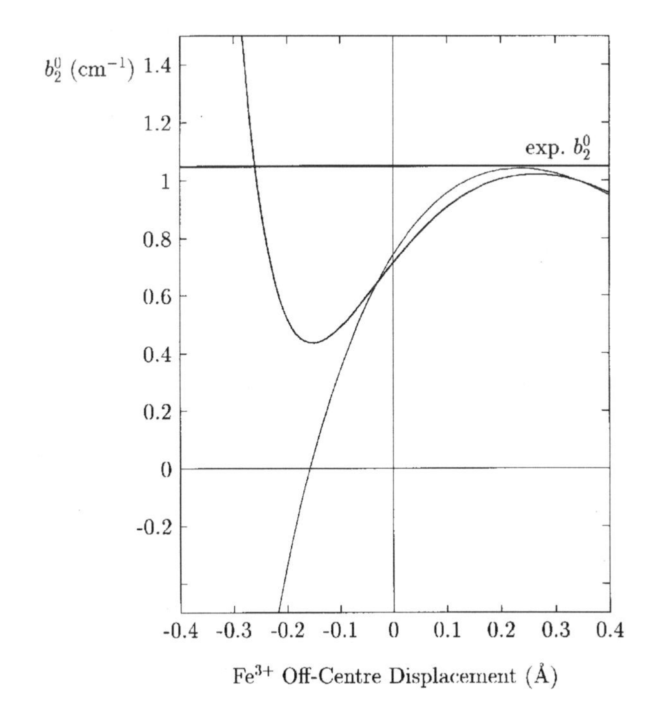

Fig. 4.12. The axial b_2^0 ESR parameter is shown as a function of the iron off-centre displacement $\delta_{\mathrm{Fe}}^{\mathrm{eff}}$. The two curves refer to the commonly employed inverse power law and to a Lennard–Jones-type dependence. It is seen that only a Lennard–Jones-type function provides a reproduction of the experimentally observed $b_2^0 \approx 1.1$ cm^{-1}consistent with an iron displacement away from the oxygen vacancy

centre [264]. The Ta site axial b_2^0 parameter is in line with former results for $Fe^{3+} - V_O$ in other perovskite crystals [245].

Zhou Yi-Yang [265] has claimed that a superposition model analysis can explain the large b_2^0 parameter in the case of Fe_K^{3+} without invoking O_I associated with the iron. Instead, he assumed a pronounced off-centre displacement of the Fe^{3+} ion combined with a considerable inward relaxation of the oxygen ligands of about 0.3 Å. Importantly, nearby interstitial oxygen has been discarded from further discussion because of its presumed negative contribution to the b_2^0 value due to the inverse power law.

In a later comment [266] it was stated that an oxygen interstitial in the vicinity of the Fe^{3+} defect is compatible with the large b_2^0 parameter if the oxygen ligands relax slightly further in the direction of the z-axis in order to outweigh the presumed negative contribution of the oxygen interstitial. However, the assumption of such large inward oxygen relaxations (> 0.3 Å) is physically not very plausible on ion-size arguments and should therefore be ruled out from explanations of the large b_2^0 parameter. After a careful re-examination of the propositions made by Zhou Yi-Yang we shall return to the idea of a nearby oxygen interstitial ion. The $Fe_K^{3+} - O_I$ defect complex can readily explain the large value of b_2^0, which was measured for the K site substitution of Fe^{3+}.

Besides defect chemical considerations (Sect. 4.1.2), shell model calculations also allow us to obtain geometrical information about the local environment of a defect. From Sect. 4.2.1 it is recalled that the isolated Fe_K^{3+}, although introducing an off-centre defect, probably behaves as a cubic ESR centre, which is due to the vanishing of b_2^0 for different iron off-centre displacements and reasonable oxygen relaxations. Referring to Fig. 4.13 this is basically due to the cancellation of the SPM contributions of the upper and central oxygen planes.

For the $Fe_K^{3+} - O_I^{2-}$ defect complex the calculations show that the oxygen interstitial moves 0.23 Å away from Fe_K^{3+}, which in turn moves 0.40 Å towards the oxygen interstitial, thus leading to a reduction of the $Fe_K^{3+} - O_I^{2-}$ distance by about 0.2 Å. The relaxed separation of the two defects is 1.83 Å. Referring to the assignment of the planes as shown in Fig. 4.13, the oxygen ligands in the lower plane (close to O_I) relax outwards, whereas ligands in the upper plane (close to the Fe^{3+}) move ≈ 0.33 Å in the direction of Fe^{3+}, resulting in an $Fe_K^{3+} - O^{2-}$ ligand separation of 2.1 Å. The distance between neighbouring oxygen ions in the upper plane becomes 2.54 Å. Ligand ions in the central plane do not move significantly. In all cases discussed the magnitude of the oxygen relaxation is compatible with generally accepted values for ionic radii.

A list of coordinates is given in Table 4.13. Because of the axial symmetry of the centre only one representative ligand ion for each plane needs to be specified.

Different from Zhou Yi-Yang [265] and Zheng Wen-Chen [266] the subsequent SPM analysis includes the Lennard–Jones-type expression. In order

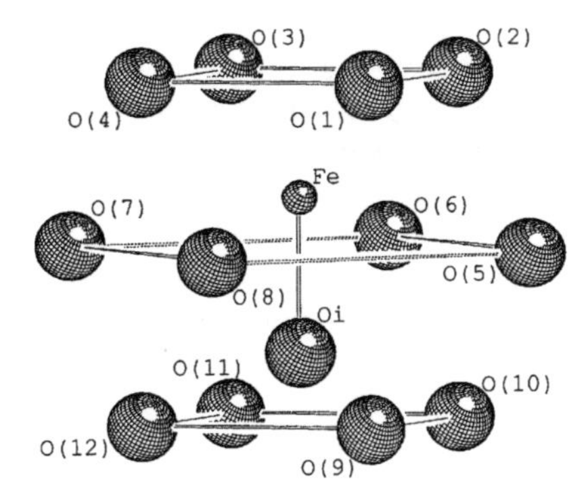

Fig. 4.13. $KTaO_3:Fe_K^{3+}–O_I^{2-}$ defect configuration. The defect is aligned on the z-axis. O(1)–O(4): upper oxygen plane; O(5)–O(8): central oxygen plane; O(9)–O(12): lower oxygen plane; O_i: oxygen interstitial

to keep the number of variables small the calculations employ a simplified treatment of the relaxation of the oxygen ligands. The 12 oxygen anions surrounding the potassium site are divided among three planes perpendicular to the axial direction of the defect centre. The displacements of the oxygen ions are confined to these planes, i.e. the movements in the upper and lower plane are along [100]-type directions and in the central plane along [110]-type directions (Fig. 4.13). The movement was parametrized by the distance of the oxygens from their original positions. Such relaxations have also been included in the approach of Zhou Yi-Yang [265]. Essentially, the results remain qualitatively unaltered when different choices of relaxation are made.

In addition, different off-centre displacements of Fe^{3+} along the z-axis are considered, including both $\bar{b}_2(r)$ dependences, i.e. the inverse power law and the Lennard–Jones-type dependence, and for isolated Fe_K^{3+} as well as

Table 4.13. Coordinates from shell model calculation for $Fe_K^{3+}–O_I^{2-}$ in $KTaO_3$ (coordinates and displacements are given in fractions of the lattice parameter a = 3.9885 Å)

Species	Unrelaxed coordinates			Relaxed coordinates			Displacement
	x_0/a	y_0/a	z_0/a	x/a	y/a	z/a	d/a
Fe^{3+}	0.0	0.0	0.25	0.0000	0.0000	0.1496	0.1004
O_I^{2-}	0.0	0.0	−0.25	0.0000	0.0000	−0.3082	0.0582
O^{2-}	0.5	0.0	0.50	0.4496	0.0000	0.4329	0.0840
O^{2-}	0.5	0.5	0.00	0.5015	0.5015	−0.0097	0.0099
O^{2-}	0.5	0.0	−0.50	0.6136	0.0000	−0.5663	0.1315

for $Fe_K^{3+} - O_I$. The local geometry is therefore given by the distance of the oxygen ions from their original positions and the off-centre coordinate of the iron defect. Figure 4.13 shows the site assumed for possible oxygen interstitial ions.

The geometrical configurations which yield the experimental value of $b_2^0 = 4.46\ \mathrm{cm}^{-1}$ are shown in Figs. 4.14 and 4.15. In the calculations on which these figures are based an oxygen interstitial is present at a fixed position 1 Å below the central oxygen plane.

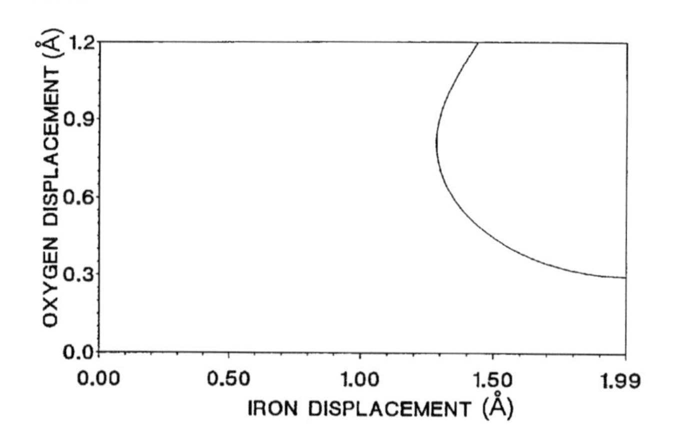

Fig. 4.14. Geometrical configurations which yield the experimental value for $b_2^0 = 4.46\ \mathrm{cm}^{-1}$ using the inverse power law $\bar{b}_2(r)$ function. An oxgen interstitial anion was positioned at 1 Å below the central oxygen plane. The iron displacement starting in the central oxygen plane is along the z-axis towards the upper oxygen plane. The oxygen ligand ions are displaced towards the z-axis within their respective planes

Figure 4.14 shows the solutions in the case of the inverse power function. In order to reproduce the measured b_2^0 value a relaxation of the oxygen ligands of at least 0.32 Å towards the z-axis is required. A displacement of the Fe^{3+} of approximately 2.00 Å along the z-axis corresponds to this oxygen displacement. Thus, the iron must approach the upper oxygen plane in order to account for the experimental b_2^0 value.

For such solutions the effect of a possible oxygen interstitial can be neglected because of its large distance from Fe^{3+}. Two important points should now be noted. Firstly, because of ion-size arguments oxygen relaxations (such as specified in this subsection) which significantly exceed 0.3 Å should be ruled out, otherwise there would be strong overlap repulsion between the oxygen ions. Already an oxygen relaxation of 0.3 Å leads to oxygen–oxygen separations of 2.4 Å (this value is to be compared with twice the oxygen radius, 2.5–2.8 Å). Secondly, in neither of the defect cases, isolated Fe_K^{3+}

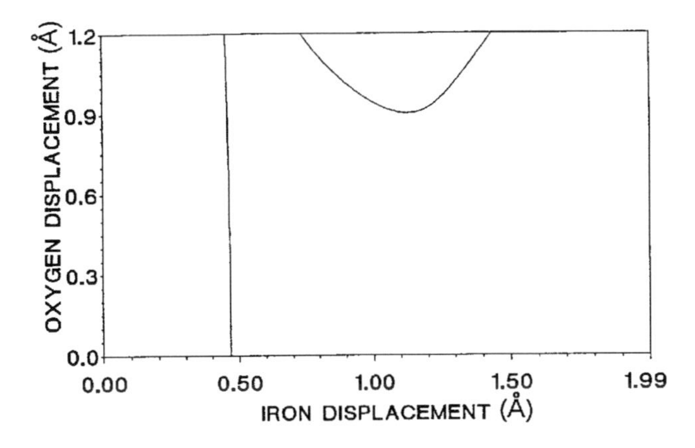

Fig. 4.15. Geometrical configurations which yield the experimental value for $b_2^0 = 4.46\ \mathrm{cm}^{-1}$ using a Lennard–Jones-type $\bar{b}_2(r)$ function. An oxygen interstitial anion was positioned at 1 Å below the central oxygen plane

or $Fe_K^{3+} - O_I$, does one find a plausible explanation for such a large Fe_K^{3+} off-centre displacement.

There are other solutions which require less off-centre displacement of the iron, but these lead to an even larger contraction of the oxygens surrounding the defect. Moreover, if corrections resulting from charge-misfit effects are included, i.e. reduction of the absolute value of $\bar{b}_2(R_0)$ by about 22%, one comes up with even more unrealistic oxygen relaxations in order to explain the observed b_2^0 value. In addition it is noted that large iron and oxygen relaxations are also ruled out on the basis of shell model calculations. In summary, the solutions in Fig. 4.14 correspond to rather unphysical relaxations.

If a superposition model analysis is performed using the Lennard–Jones function instead of the inverse power law, one observes that the $Fe^{3+} - O_I$ defect complex together with moderate oxygen ligand relaxations is the only physically reasonable solution. The straight line in Fig. 4.15 corresponds to this solution. The curved part in Fig. 4.15 belongs to solutions based on large oxygen ligand relaxations, which are physically unreasonable. Calculations with the Lennard–Jones-type radial function yield the correct b_2^0 value for a $Fe_K^{3+} - O_I$ distance of ≈ 1.5 Å. This can clearly be seen from Fig. 4.15, where the line indicates the correct solution for an iron off-centre shift of just under 0.5 Å. To this the distance of the interstitial oxygen from the central oxygen plane of 1.0 Å (fixed) has to be added. Shifting the position of the oxygen interstitial results in a parallel shift of the straight line in Fig. 4.15, leading to the same distance between $Fe_K^{3+} - O_I$ of ≈ 1.5 Å in every case.

It is emphasized that the superposition model analysis given so far can only give some useful qualitative hints, because of the well known problems with $\bar{b}_2(r)$ functions, and because of the possible limitations of the superposition model itself in this special case.

Thus, we should not attempt to interpret the calculated $\mathrm{Fe}_{\mathrm{K}}^{3+}-\mathrm{O}_{\mathrm{I}}$ distance of 1.5 Å as being significant, but instead we should invoke the principle occurrence of a physically satisfactory solution if we are using the Lennard–Jones-type function. It is finally recalled that the occurrence of this axial K site defect complex originates in the combined effect of self-compensation of iron and an iron-assisted oxygen Frenkel defect formation (Sect. 4.1.2).

5. Lithium Niobate

When considering photorefractive oxides, most attention has been devoted to $LiNbO_3$. The technical properties largely depend on the type and concentration of defects. Whereas Ti-doped $LiNbO_3$ is used as a waveguiding medium, the incorporation of iron makes it useful for photorefractive applications. In contrast with all other oxide materials discussed in this volume, the defect chemistry of $LiNbO_3$ and relevant material properties are largely determined by high concentrations (in the percentage range) of cationic intrinsic defects, which are formed to accommodate the most commonly observed Li_2O non-stoichiometry. The precise knowledge of these intrinsic defect structures is indispensable for a microscopic understanding of impurity-related effects.

The photorefractive properties of Li_2O-deficient $LiNbO_3$ are subject to drastic changes upon doping this material with magnesium. Whereas the photorefractive effect is useful for the storage of volume phase holograms, it is unwanted in optical waveguides, where it induces disturbing scattering effects. Zhong et al. [267] discovered that this optical damage is significantly reduced in crystals grown from a congruent melt containing at least 4.6 mol-% MgO[1]. Several physical properties show characteristic changes when the MgO content is raised above this threshold concentration. Analogous effects are observed by keeping the MgO content fixed and varying instead the [Li]/[Nb] ratio [268, 269]. Thus Mg cations only indirectly influence the threshold behaviour. The key to an understanding of the underlying physics relates to the idea that the magnesium solution modes are highly correlated with the stoichiometry-induced intrinsic defect structures in $LiNbO_3$.

The following sections summarize the most important results of electronic-structure and shell model simulations. In particular, the latter modelling studies provided a wealth of information on the ionic aspects of defect formations in $LiNbO_3$.

5.1 Electronic Structure Calculations

Ferroelectric $LiNbO_3$ belongs to the family of perovskite oxides. However, the small size of Li ions leads to significant distortions of the crystal struc-

[1] In comparison, the crystalline MgO concentration is enhanced by a factor 1.2, i.e. $[MgO]_{cryst} \sim 5$–6 mol-%.

ture with respect to the ideal perovskite structure (see [270] for details). The paraelectric space group is $R\bar{3}c$ and the ferroelectric $R3c$. The complicated structure might be the reason that there are only two band structure calculations for this material [52, 271]. There are no efforts to elucidate the nature of ferroelectricity by employing *ab initio* calculations. However, the FE-PT driving mechanisms could be similar to those discussed for oxide perovskites in Sect. 3.1.

The recent calculations of Ching et al. [52] are of first-principle-type. They are based on an LCAO scheme within DFT and assume perfectly built stoichiometric $LiNbO_3$. Further, the LDA approximation has been improved by introducing a quasi-particle self-energy correction. As a result, the pure LDA gap of 2.62 eV increases to 3.56 eV in the self-energy corrected case. In comparison the experimental optical gap equals 3.78 eV. Moreover, the calculated gaps are indirect along $\Gamma - X$ with energy differences corresponding to a few hundredths of an eV. All subsequently reported results refer to the self-energy corrected calculation. The upper valence band (VB) has a width of 5.3 eV and corresponds mainly to oxygen 2p-states (with covalent admixtures of Nb 4d states); the lower conduction band (CB), in turn, essentially consists of Nb 4d levels. The CB width is 2 eV. An analysis of the charge density map supports the idea of an ionic crystal with covalent bonding contributions between niobium and oxygen. The authors report the formula $Li^{0.98}Nb^{3.67}(O^{-1.55})_3$. Ching et al. also calculated the dielectric function $\varepsilon(\omega)$ as well as the ordinary and extraordinary refractive indices $n_o(\omega)$, $n_e(\omega)$. Most importantly, the results show that the inclusion of a self-energy correction is necessary in order to obtain reliable values. The reported excellent agreement between theoretical and experimental refractive indices seems to be fortuitous to some extent, because the experimental values refer to non-stoichiometric crystals. Theoretical refractive indices are by 0.1–0.2 too large. However, the agreement is satisfactory in comparison to uncorrected LDA results.

Up to the present day only a few studies have been concerned with the electronic structure of defects. So far, all investigations have employed a simplified cluster approach. The earlier calculations of Michel-Calendini et al. [272, 273] were based on Slater's Xα approximation and on the Muffin Tin potential approximation. The clusters consisting of a central cation M^{n+} (Nb^{5+}, $Fe^{2+/3+}$ or Cr^{3+}) and its nearest oxygen neighbours were embedded inside a neutralizing Watson sphere. Additionally, in the case of M_{Li}^{n+} a larger cluster corresponding to $(NbMO_9)^{(13-n)-}$ has been considered. The major inaccuracies of these calculations might be related to the Watson sphere (see Sect. 2.1), to the MT approximation and to the neglect of lattice relaxations. Care is needed when interpreting calculated energies. Most reliable are total energies or orbital energy differences calculated on the basis of the transition state procedure [18]; otherwise, orbital energies have no proper meaning, because Koopmans' theorem does not apply to the Xα scheme. Further, the MT approximation has been stated [78] to underestimate the covalency effects in favour of an atomically localized description influencing crystal field splittings and energy level positions. From such calculations one can certainly draw valuable qualitative conclusions concerning defect levels and optical properties. However, they are insufficient to decide upon the incorporation site of impurities. To indicate computational implications I quote a further cluster calculation on Cr^{3+} in $LiNbO_3$ which also employed the $X\alpha$ approximation but did not use Muffin Tin potentials [274]. These calculations have been done *in vacuo*, i.e. all effects of the embedding crystal lattice have been neglected. Further, a frozen core approximation has been applied to oxygen 1s states and Cr 1s-3p states. More importantly, oxygen-related polarizing d-functions, which are known to be necessary to model oxygen ions properly, have been omitted from the basis set used. As a result the calculated optical transition energies between crystal-field states are significantly reduced for Cr_{Nb}^{3+} and increased in the case of Cr_{Li}^{3+}, compared with the results of Michel-Calendini. In spite of the similarity of the cluster models the specific computational differences lead to deviations in crystal field energies ranging between 2000 and 6000 cm^{-1}.

DeLeo et al. [275] employed $X\alpha$ cluster calculations in order to study a hypothetical oxygen vacancy. The calculations were done for a small and a large cluster, i.e. for $V_OLi_2Nb_2$ and $V_OLi_2Nb_2O_{12}$, respectively. The calculations employed embedding Watson spheres. In the case of the large cluster, additional corrections due to the crystalline Madelung potential were taken into account. Without discussing this work in further detail, it is worth emphasizing that an electron trapped by the vacancy (F^+ centre) becomes localized on the Nb ions adjacent to the vacancy. There are hints that this behaviour of oxygen vacancies is characteristic for all oxide perovskites (see also Chap. 3).

The recent *in vacuo* cluster calculations of Michel-Calendini et al. [276, 277] employ the linear combination of Gaussian-type basis functions within DFT-LDA. Total energy calculations have been used to study ground and excited crystal field states of 3d transition metal cations in $LiNbO_3$. Though there is no general justification for excited state DFT these investigations may be correct for the following reason. All considered states have different symmetries. Therefore, the excited states can be interpreted as being ground states within their respective symmetry classes; it is, however, important to implement appropriate symmetry restrictions during density optimization [278]. Without imposing any symmetry restrictions the investigations correspond to excited state calculations using the ground state energy functional, leading to lower bounds for the required crystal field splittings. This argument complies with the theorem of Perdew and Levy, which was considered briefly in Sect. 2.1. The calculations of Michel-Calendini et al. demonstrate the sensitivity of energy differences as a function of the cation positions along the *c*-axis. However, contrary to the authors' claim, lattice site occupations cannot be predicted on the basis of these *in vacuo* cluster calculations due to the neglect of all the important effects of the embedding host lattice which, for example, controls the donor/acceptor behaviour of impurities on the Li or Nb site, respectively.

5.2 Shell Model Simulations of Defect Chemical Properties

The ionic picture emerging from the band structure calculations of Ching et al. (Sect. 5.1) backs the earlier simulations of Donnerberg et al. [124, 279] which were based on a formally ionic shell model description. These simulations successfully reproduced both the $R\bar{3}c$ and the ferroelectric $R3c$ phases. Potential parameters were either transferred from appropriate binary oxides or were obtained through empirical fitting procedures with respect to the structural, elastic and dielectric properties of $LiNbO_3$ (see also Sect. 2.2.2). In particular, the inclusion of the oxygen core–shell polarization vectors as adjustable variables during the fitting facilitated the simulation of the ferroelectric phase, being about 0.1 eV more favourable than the paraelectric crystal phase.

The major interest in these potential simulations has been devoted to elucidating the defect chemistry of $LiNbO_3$. The most important results [124, 279, 280] are summarized in the following subsections. Note that many of the impurity-related predictions have been *a posteriori* confirmed by ESR, EXAFS and channelling experiments. However, it should be emphasized that the interpretation of channelling-induced dips [281] in particular refers to the observed perfect crystal structure of $LiNbO_3$. Deviations due to imperfections and lattice distortions are not accounted for systematically. It is not clear to what extent this neglect could be responsible for certain inconsistencies. As an example we shall consider the magnesium solution in $LiNbO_3$ (see Sect. 5.2.2).

5.2.1 Intrinsic Defect Structure

The intrinsic defect structure is dominated by defects related to Li_2O deficiency, which is found to exist in most crystal samples. Non-stoichiometry is accommodated on the cationic sublattices, whereas the oxygen sublattice remains unaffected. The energetically most favourable reaction (in Kröger–Vink notation [205]) is given by

$$'\mathrm{LiNbO_3}' \longrightarrow 3\mathrm{Li_2O} + 4\mathrm{V}^{\mathrm{l}}_{\mathrm{Li}} + \mathrm{Nb}^{4\bullet}_{\mathrm{Li}} - \mathrm{LiNbO_3}\,. \tag{5.1}$$

Equation (5.1) describes the out-diffusion of Li_2O from the bulk of $LiNbO_3$ (denoted as $'LiNbO_3'$). The formation of cationic defects is accomplished by dissolving one formula unit $LiNbO_3$ from the surface into the bulk. For congruently grown $LiNbO_3$ (~48.5 mol-% Li_2O) (5.1) predicts about 1 mol-% Nb-antisite defects. On the other hand, earlier X-ray studies of Abrahams and Marsh [282] and NMR investigations of Peterson and Carnevale [283] suggested the formation of 5–6 mol-% Nb antisites according to the reaction

$$'LiNbO_3' \longrightarrow 3Li_2O + 4V_{Nb}^{5l} + 5Nb_{Li}^{4\bullet} - LiNbO_3\,, \tag{5.2}$$

which may be rearranged to become

$$'LiNbO_3' \longrightarrow 3Li_2O + 4(V_{Nb}^{5l}Nb_{Li}^{4\bullet}) + Nb_{Li}^{4\bullet} - LiNbO_3\,. \tag{5.3}$$

Equation (5.3) looks similar to (5.1) with $(V_{Nb}^{5l}Nb_{Li}^{4\bullet})$ replacing the ordinary Li vacancies. Model (5.1) corresponds to the chemical sum formula $Li_{1-5x}Nb_{1+x}O_3$ and the models (5.2–5.3) to $(Li_{1-5x}Nb_{5x})Nb_{1-4x}O_3$.

Shell model calculations have shown that (5.3) is by 10.6 eV per Li_2O molecule less favourable than reaction (5.1), and does not represent a realistic alternative to (5.1). Smyth [284] resolved this contradictory situation by the observation that (5.2) is equivalent to (5.1) if one assumes the Li_2O deficiency to occur within ilmenite-structured $LiNbO_3$ intergrowths. Perfect $LiNbO_3$ and ilmenite-like $LiNbO_3$ differ in their cationic stacking sequence along the c-axis, i.e.

$$...NbLi\Box NbLi\Box NbLi\Box NbLi... \tag{5.4}$$

$$...NbLi\Box LiNb\Box NbLi\Box LiNb...\,, \tag{5.5}$$

respectively, with □ denoting structural cation vacancies. Therefore the ilmenite structure offers a natural source of additional Nb antisites. Both structural modifications of $LiNbO_3$ are energetically almost equivalent. The calculated lattice energies differ by 0.1 eV per formula unit in favour of perfect $LiNbO_3$, which is in excellent agreement with the calorimetric measurements of Mehta et al. [285]. Further, the loss of Li_2O has been found to be almost equivalent energetically within both structural modifications [124]. Together, these results are able to lend support in favour of the Smyth model.

Due to the observation that 5–6 mol-% Nb antisites correspond exactly to the Li_2O deficiency in congruent $LiNbO_3$ according to (5.2), one must assume that the hypothetical ilmenite-structured intergrowths are massively Li-depopulated. Additional ilmenite-structured but stoichiometric crystal regions would lead to surplus antisite defects being independent of (5.2) and, therefore, the total antisite concentration should significantly exceed the observed 5–6 mol-%. Stated in a different way, the present discussion suggests the formation of extended clusters consisting of intrinsic stoichiometry-related defects, and the structure of these clusters resembles the ilmenite modification.

However, the structure investigations of Iyi et al. [286] and Wilkinson et al. [287] based on sensitive neutron diffraction techniques could not reproduce the results of Abrahams and Marsh, but favoured the Li vacancy model (5.1). Moreover, the NMR studies of Blümel et al. [288] are in agreement with these recent structural analyses. Therefore one is inclined to favour the Li vacancy model (5.1); however, the ultimate proof is still lacking. It does not seem unambiguously clear which of the two models correctly describes the intrinsic defect chemistry of $LiNbO_3$ related to the observed non-stoichiometry. It might be possible that both models are correct to some extent, since the detailed growth conditions could influence nature's choice of the particular reaction. Additional investigation must be devoted to this topic.

In view of this situation, both model options, which merely differ by a shift of Nb cations, should be considered. Table 5.1 displays the binding energies of some basic intrinsic defect clusters. Two important conclusions may be drawn from Table 5.1. First, it is seen that the energy gain due to agglomeration is more efficient in the Nb vacancy model of Abrahams and March which complies with the preceding discussion of ilmenite-structured $LiNbO_3$ representing an infinite cluster description. Second, significant binding energies are also obtained in the ordinary Li vacancy model. Thus, both defect models indicate the existence of intrinsic defect clusters. For stability reasons these clusters may be imagined as consisting of the electrically neutral building units $4V^{l}_{Li}$ ... $Nb^{4\bullet}_{Li}$ or $4(Nb^{4\bullet}_{Li}V^{5l}_{Nb})$ ... $Nb^{4\bullet}_{Li}$. In particular, the discussion of the Nb vacancy model suggests that these clusters should be spaciously extended in order to be competitive energetically with the propper Li vacancy model. It is noted that, in principle, the Li vacancy model is also compatible with extended defect clusters.

Table 5.1. Binding energies between intrinsic defects. The energies refer to defect species separated by "..."

Defect species	Binding energy (eV)
$(Nb^{4\bullet}_{Li}V^{5l}_{Nb})$ ... $Nb^{4\bullet}_{Li}$	−2.4
V^{l}_{Li} ... $Nb^{4\bullet}_{Li}$	−0.7
V^{l}_{Li} ... $(3V^{l}_{Li}Nb^{4\bullet}_{Li})$	−0.3

Unfortunately, Mott–Littleton-type defect calculations (see Sect. 2.2) are not feasible for investigating the energetics of such extended defect clusters, but there are independent experimental observations which may be interpreted in favour of stoichiometry-related defect clusters. The observed inhomogeneity of $LiNbO_3$ single crystals detected by TEM measurements [289] seems to agree with the proposition of extended intrinsic defect clusters. Support is also given by electron spin resonance (ESR) investigations of Nb^{4+} in electrochemically reduced $LiNbO_3$ (see also at the end of this subsection). The defects could be assigned to $Nb^{3\bullet}_{Li}$ antisites with associated defects on neighbouring cation sites [290] (it is emphasized that the ESR signal of normal-site small polarons Nb^{l}_{Nb} looks significantly different [133]). Further, measurements of the AC response of reduced congruent $LiNbO_3$ proved these crystals to be highly inhomogeneous [291]. The observed partition into insulating and conducting regions might be understood on the basis of extended stoichiometry-related defect clusters which contain Nb-antisite defects acting as electron traps. Therefore the defect clusters would constitute the AC conducting regions.

Both non-stoichiometry defect models can be unified by employing the generalized notation V^{l} to denote the singly charged cation vacancies, which are part of the assumed defect clusters. From a structural point of view these vacancies may resemble either ordinary Li vacancies or the defect complexes $(\text{Nb}_{\text{Li}}^{4\bullet}\text{V}_{\text{Nb}}^{5\text{l}})$.

The calculations on Li vacancies bound to antisites, reported in Table 5.1, indicate that the binding energy of a cation vacancy is a function of the local ratio of concentrations $[\text{V}^{\text{l}}]/[\text{Nb}_{\text{Li}}^{4\bullet}]$. Thus, one may assume that the following relation holds true:

$$E(\text{V}^{\text{l}}) = \begin{cases} > E(\text{V}_{\text{Li}}^{\text{l}}) \text{ for } [\text{V}^{\text{l}}]/[\text{Nb}_{\text{Li}}^{4\bullet}] > 4 \\ < E(\text{V}_{\text{Li}}^{\text{l}}) \text{ for } [\text{V}^{\text{l}}]/[\text{Nb}_{\text{Li}}^{4\bullet}] \leq 4 \,. \end{cases} \tag{5.6}$$

Whereas $E(\text{V}^{\text{l}})$ refers to cation vacancies within defect clusters, $E(\text{V}_{\text{Li}}^{\text{l}})$ denotes the formation energy of isolated Li vacancies. It is emphasized that an analogous relation does not make sense for isolated intrinsic defects. Instead, the defects should be considered as being completely independent of each other. The defect chemical consequences of these pronounced model differences will be discussed further in the context of impurity solution modes (Sect. 5.2.2).

Analogous to non-stoichiometry, electrochemical reduction of LiNbO_3 affects only the cationic sublattices. In particular the shell model calculations give no hints in favour of oxygen vacancies. The following reduction mechanisms are favourable:

$$'\text{LiNbO}_3' \longrightarrow \text{O}_2 + \text{Li}_2\text{O} + \text{Nb}_{\text{Li}}^{4\bullet} + 4\text{e}^{\text{l}} - \text{LiNbO}_3 \,, \tag{5.7}$$

$$'\text{LiNbO}_3' + 2\text{V}^{\text{l}} \longrightarrow \frac{3}{2}\text{O}_2 + \text{Nb}_{\text{Li}}^{4\bullet} + 6\text{e}^{\text{l}} - \text{LiNbO}_3 \,. \tag{5.8}$$

Reaction (5.8) takes advantage of stoichiometry-related cation vacancies. e^{l} denotes $\text{Nb}_{\text{Nb}}^{4+}$ small polarons which are generated during reduction. Electron spin resonance proves that the electronic ground state of LiNbO_3 after reduction is diamagnetic. Isolated polarons $(\text{Nb}_{\text{Li}}^{4+})$ are observable only after illumination or thermal activation. These results are in agreement with shell model-based simulations which suggest that the formation of electronic bipolarons is favourable. The bipolarons consist of Nb pairs $(\text{Nb}_{\text{Li}}^{4+} - \text{Nb}_{\text{Nb}}^{4+})$ orientated along the c-axis. Bipolarons are stabilized by lattice relaxations and covalency. Crude estimates of direct covalent bonding contributions between the respective niobium $4\text{d}(\text{e}_{\text{g}})$-type orbitals may be given on the basis Harrison's solid state matrix elements [20]. The resulting binding energies have been reported to be of the order of 1 eV [124].

Finally, we briefly consider a recent crystal growth technique which allows us to obtain stoichiometric $LiNbO_3$ crystals of high quality. These samples show reduced optical damage effects due to the absence of intrinsic defects. The growth technique employs a congruent Li_2O–Nb_2O_5 melt which is enriched by adding the oxide K_2O ($\sim$ 10 mol-% [292]). Note that potassium does not enter the $LiNbO_3$ crystals significantly. Possible explanations of these results can be based on the the lowering of the melting temperature of the mixed system Li_2O–K_2O–Nb_2O_5 facilitating the formation of stoichiometric $LiNbO_3$. Further arguments may be derived from simplified shell model simulations: due to their similar lattice energies both $LiNbO_3$ and $KNbO_3$ may be assumed to be grown from the melt. Shell model simulations yield $E_{\text{Lat}} = -173.8$ eV and $E_{\text{Lat}} = -173.3$ eV per formula unit of $LiNbO_3$ and $KNbO_3$, respectively. Since K_2O tends to bind a significant proportion of Nb_2O_5, the melt ratio $[Li_2O]/[Nb_2O_5]$ is effectively increased, which in turn favours the formation of stoichiometric $LiNbO_3$. Moreover, the Li/K cation exchange (solid state) reaction

$$'\mathrm{KNbO_3}' +' \mathrm{LiNbO_3}' \longrightarrow \mathrm{Li}_{\mathrm{K}}^{\times} + \mathrm{K}_{\mathrm{Li}}^{\times} \tag{5.9}$$

needs 4.4 eV per cation pair, which clearly indicates that the system avoids any mixing of $LiNbO_3$ and $KNbO_3$. For simplicity we have considered the formation of $KNbO_3$ during the process, but using instead any different compound of the K_2O–Nb_2O_5 subsystem would not change the quality of the arguments.

5.2.2 Incorporation of Impurities

The incorporation sites for impurity cations and the possible modes of charge compensation have been found to depend on the degree of the Li_2O non-stoichiometry. The impurity concentrations given below refer to crystalline values, not to melt compositions. They are defined in terms of ternary phase diagrams made up of Li_2O, Nb_2O_5 and the actual impurity oxide, i.e.

$$[\mathrm{M}_n\mathrm{O}_m] = \frac{N(\mathrm{M}_n\mathrm{O}_m)}{N(\mathrm{Li_2O}) + N(\mathrm{Nb_2O_5}) + N(\mathrm{M}_n\mathrm{O}_m)}\,, \tag{5.10}$$

where N denotes the number oxide molecules given in brackets.

First we briefly discuss stoichiometric crystals which are characterized by the absence of Nb antisites and cation-type vacancies. It is noticed that congruently grown crystals containing high concentrations of appropriate impurity cations (e.g. 4.6 mol-% magnesium in the melt) behave similar to stoichiometric $LiNbO_3$, in that massive doping leads to the disappearance of stoichiometry-induced intrinsic defects. If additional impurities are to be solved, the respective solution modes correspond to stoichiometric $LiNbO_3$. As an example, we consider divalent impurity cations M^{2+}. A choice of possible incorporation reactions is given by

$$\mathrm{MO} +' \mathrm{LiNbO_3}' \longrightarrow \mathrm{M_{Li}^{\bullet}} + \mathrm{V_{Li}^{l}} + \mathrm{Li_2O}\,, \tag{5.11}$$

$$\mathrm{MO} +' \mathrm{LiNbO_3}' \longrightarrow \mathrm{M_{Li}^{\bullet}} + \mathrm{e^{l}} + \frac{1}{4}\mathrm{O_2} + \frac{1}{2}\mathrm{Li_2O}\,, \tag{5.12}$$

$$2\mathrm{MO} +' \mathrm{LiNbO_3}' \longrightarrow 2\mathrm{M_{Nb}^{3l}} + 3\mathrm{V_{O}^{\bullet\bullet}} + \mathrm{Nb_2O_5}\,, \tag{5.13}$$

$$4\mathrm{MO} +' \mathrm{LiNbO_3}' \longrightarrow 3\mathrm{M_{Li}^{\bullet}} + \mathrm{M_{Nb}^{3l}} + \mathrm{LiNbO_3} + \mathrm{Li_2O}\,. \tag{5.14}$$

Similar equations may be formulated for cations with different charge states. Equation (5.14), i.e. self-compensation, represents the most favourable solution mode for all divalent cations. Self-compensation is also preferred for other impurity charge states [279]. Various experimental data confirm the occupation of Li and Nb sites. ESR measurements on $LiNbO_3$:Mg co-doped with iron ([Mg] > 5 mol-%) demonstrated the existence of two different Fe^{3+} signals [293]. The given interpretation due to Li and Nb site occupation has been supported by recent extended X-ray absorption fine structure (EXAFS) investigations on $LiNbO_3$:Mg,Fe [294]. Malovichko et al. [295] identified Fe_{Li}^{3+} and Fe_{Nb}^{3+} defects in otherwise stoichiometric $LiNbO_3$. Further evidence is given by the incorporation of magnesium: for very large MgO concentrations reaction (5.14) directly explains the formation of $Mg_4Nb_2O_9$ which is well known from the Li_2O–MgO–Nb_2O_5 phase diagram [296]. It is finally recalled that self-compensation has been also predicted to be favourable in the related $A^{1+}B^{5+}O_3$ (Chap. 4) and $A^{2+}B^{4+}O_3$ (Chap. 3) perovskites and in SBN (Chap. 6). A further phenomenon which is closely related to self-compensation refers to the intrinsic cation interchange (or inversion) in YIG (Sect. 7.3) and SBN (Chap. 6), it is also known to occur in many spinel-type oxides AB_2O_4 [12]. As a common feature we may infer the incorporation of a particular cation species at different sites in one crystal.

Within a few tenths of an eV (5.11) and (5.12) are energetically competitive to each other, but external reducing conditions would be in favour of (5.12). Due to the factor 1/4 the formation entropy of gaseous oxygen does not appreciably affect the calculated reaction energies. In any case, both reactions are by ~ 2 eV significantly less favourable than self-compensation. It is noted, however, that electronic compensation could receive some limited relevance due to deep electronic gap levels. This possibility has also been discussed for the other ABO_3 oxide materials (see Sects. 3.2 and 4.1.2).

In order to circumvent the creation of both donor and acceptor defects as induced by self-compensation it is worth considering the incorporation of hexavalent impurity cations M^{6+}, e.g. tungsten. Certainly, these cations fit best in the Nb sites, thus forming donor-type defects. The most favourable solution modes are given by

$$\mathrm{MO_3} +' \mathrm{LiNbO_3}' \longrightarrow M_{Nb}^{\bullet} + V_{Li}^{l} + \mathrm{LiNbO_3}\,, \tag{5.15}$$

$$\mathrm{MO_3} +' \mathrm{LiNbO_3}' \longrightarrow M_{Nb}^{\bullet} + e^{l} + \frac{1}{4}O_2 + \frac{1}{2}\mathrm{Nb_2O_5}\,, \tag{5.16}$$

with the reaction energies being equal up to 0.1 eV. The earlier remarks on (5.11) and (5.12) also apply to the present case. One may thus assume that electrons and Li vacancies will compensate for the solution of hexavalent impurity cations.

It is important to note that the impurity incorporation modes in stoichiometric $LiNbO_3$ do not impose any limits for solubility. This situation is completely different in Li_2O-deficient $LiNbO_3$, to which we now turn.

For Li_2O-deficient crystals most cations are incorporated at Li sites irrespective of their charge state. Charge compensations are maintained through consumption of intrinsic Nb antisites and cation vacancies. For example, the most favourable solution mode for low concentrations of divalent cations M^{2+} is given by (see also [279, 280])

$$5MO + 3V^{l} + 2Nb_{Li}^{4\bullet} \longrightarrow 5M_{Li}^{\bullet} + Nb_2O_5 \,, \qquad (5.17)$$

with the understanding that V^{l} refers either to Li-type vacancies or $(Nb_{Li}^{4\bullet}V_{Nb}^{5l})$ (see Sect. 5.2.1). In what follows the detailed structure of the included cation vacancies is not important. Only the assumed extensive agglomeration of cation vacancies and Nb antisites will play a significant role. But irrespective of the defect cluster hypothesis, all impurity solution modes in Li_2O-deficient $LiNbO_3$ are subject to particular solubility limits which are determined by the finite amount of V^{l}-compensating Nb antisite defects (i.e. 1 mol-% of the total niobium content in congruent $LiNbO_3$). For example, the solubility limit in (5.17) is given by $[MO]_{max} = 2.5$ mol-%. It is important to note that the Nb antisite defects exactly disappear at the respective solubility limit. Moreover, if the stoichiometry-induced defects were not bound to clusters, solution modes like (5.17) would be operative up to their particular solubility limits, giving rise to a change of the incorporation mechanism at even higher impurity concentrations.

To be explicit we next consider the incorporation of magnesium Mg^{2+}. Figure 5.1 displays the calculated [279, 280] and measured Li_2O and Nb_2O_5 concentrations as a function of the MgO content. The experimental data have been taken from chemical analyses performed by Grabmaier et al. [297]). Recent investigations of Iyi et al. [298] strongly support these earlier data. It is observed from Fig. 5.1 that the Li_2O concentration is significantly more affected by the MgO content than the Nb_2O_5 counterpart. The experimentally established dependence $[Li_2O]=f([MgO])$ suggests the existence of two pronounced kinks, with the first occurring at $\sim$1.5 mol-% magnesium and the second between 5 and 6 mol-% MgO (i.e. the well-known threshold concentration).

Any theoretical model accounting for these two kinks must be, in addition, consistent with further experimental constraints. ESR and optical absorption experiments on Mg-doped $LiNbO_3$ proved the existence of Nb antisites for all concentrations below the threshold value, but antisites cease to exist above the threshold [299, 300]. This information complies with second harmonic generation (SHG) investigations [301]. The composition dependence of the SHG phase-matching temperature is significantly affected in vapour transport equilibrated (VTE) materials for all Mg concentrations below the threshold.

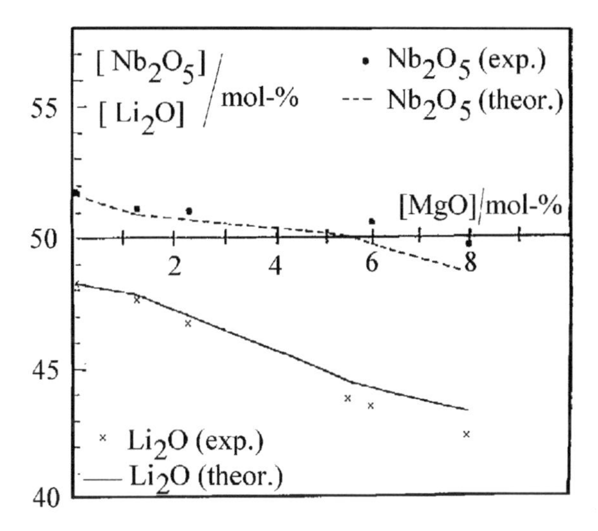

Fig. 5.1. Comparison of calculated and experimental Li_2O and Nb_2O_5 concentrations as a function of the MgO content. The results refer to congruent melt compositions

There are no corresponding changes above this critical concentration. Thus, VTE treatments only influence Li_2O-deficient $LiNbO_3$ containing antisites and cation vacancies.

One might attempt to explain the threshold concentration due to the reaction

$$5MO + 5Li_{Li}^{\times} + Nb_{Li}^{4\bullet} \longrightarrow 5M_{Li}^{\bullet} + V_{Li}^{l} + LiNbO_3 + 2Li_2O\,, \tag{5.18}$$

neglecting the fact that it is not the most favourable solution mode for small impurity concentrations, but which could be operative for all concentrations below the threshold value of 5 mol-%. However, this reaction, when taken alone, is unable to reproduce the observed dependence $[Li_2O]=f([MgO])$ in this regime. In particular, no account is made to model the first kink.

In general, any model is to be ruled out which aims to relate the first kink to a solubility limit due to disappearence of antisites. Corresponding models have been recently suggested by Iyi et al. [298] and by Kling et al. [281]. In both of these scenarios Nb antisites disappear at about 3 mol-% magnesium. It is emphasized that this value is twice as large as the observed concentration of the first kink shown in Fig. 5.1. The model of Kling et al., which is based on interpretations of channelling results, may be written as:

$$3MgO + V^{l} + Nb_{Li}^{4\bullet} + Li_{Li}^{\times} \longrightarrow 3Mg_{Li}^{\bullet} + LiNbO_3 \quad \text{for } [MgO] < 3\text{ mol-\%} \tag{5.19}$$

and

$$3\mathrm{MgO} + \mathrm{V}^{\mathrm{l}} + \mathrm{Nb}_{\mathrm{Nb}}^{\times} + \mathrm{Li}_{\mathrm{Li}}^{\times} \longrightarrow 2\mathrm{Mg}_{\mathrm{Li}}^{\bullet} + \mathrm{Mg}_{\mathrm{Nb}}^{3\mathrm{l}} + \mathrm{LiNbO}_3 \quad \text{for } [\mathrm{MgO}] > 3\ \text{mol-\%}\,. \tag{5.20}$$

Moreover, this model suffers from its inability to provide an explanation for the threshold phenomena around 5 mol-%. It is recalled that explanations of channelling experiments employ the perfect crystal structure, which could lead to misinterpretations for highly perturbed $LiNbO_3$. The model of Iyi et al. extends the channelling-based model by inclusion of the solution reaction (5.11) in the intermediate regime 3 mol-% < [MgO] < 5 mol-%, with the self-compensation-type reaction (5.20) then being employed beyond the threshold. Although able to explain the existence of two kinks, this latter model bears the additional inconsistency that (5.11) is by 2.6 eV less favourable than (5.20).

It is important to realize that the occurrence of the first kink cannot be reasonably accounted for on the basis of isolated defect models, since any attempts at explanation would rest on solubility limit arguments. Therefore, the observed kink at about 1.5 mol-% magnesium, which has been also identified in Zn-doped crystals [302], directly indicates the existence of extended intrinsic defect clusters, as proposed in the preceding section.

The calculated concentration dependences displayed in Fig. 5.1 anticipate the existence of extended intrinsic defect clusters. It is assumed that magnesium cations substitute for the vacancies and Nb antisites forming these clusters. On the basis of (5.17) the (divalent) Mg ions predominantly replace the (pentavalent) antisites at low dopant concentrations, leading to an increase of the ratio $[\mathrm{V}^{\mathrm{l}}]/[\mathrm{Nb}_{\mathrm{Li}}^{4\bullet}]$ and, thus, to a destabilization of the cation vacancies (see (5.6)). Consequently, the kink around 1.5 mol-% refers to a crossover to solution modes compensating for the rising imbalance: Mg cations still replace both types of stoichiometry-induced defects, but the emphasis is now placed on substituting for cation vacancies. The corresponding solution modes, which are chosen to be as energetically favourable as possible, require the formation of additional charge-compensating intrinsic defects, of which Li vacancies represent the most natural choice in this case[2]. The model is explained in some detail in [280]. Approaching the threshold concentration the stoichiometry-induced defects disappear completely, and the relevant solution mode becomes (5.20) beyond the threshold. Finally, after consumption of any Li vacancies, proper self-compensation (5.14) substitutes for reaction (5.20), leading to inclusions of the complex compound $Mg_4Nb_2O_9$ for magnesium concentrations greater than 25 mol-% [296].

[2] Electronic compensation would be almost equally favourable, but could not account for the observed decrease of the Li_2O-content, i.e. $d[\mathrm{Li_2O}]/d[\mathrm{MgO}] \approx -1$ between 1.5 and 5–6 mol-% MgO. Instead, only oxygen gas would be released into the environment. Further, the typical ESR and optical absorption spectra should be observable. However, this has not been reported. In summary, although electronic compensation cannot be completely ruled out on theoretical grounds, this is finally possible due to the experimental facts.

The solution of trivalent impurities behaves differently from the divalent counterpart discussed above. For trivalent cations (Sc^{3+} and In^{3+}) the threshold concentration is reduced to 1–2 mol-% [300, 303]. In this case the threshold concentration may correspond to the solubility limit (1.67 mol-%) of the predicted most favourable solution mode [279]:

$$M_2O_3 + 4V^{l} + 6Nb_{Li}^{4\bullet} \longrightarrow 10M_{Li}^{\bullet\bullet} + 3Nb_2O_5 \,. \tag{5.21}$$

By assuming intrinsic defect clusters, this means that trivalent cations are better suited to substituting for antisites than divalent ions, in that the impurity-induced charge misfit perturbations of the defect cluster are reduced. It is further noted that trivalent cations are more effective in consuming the antisites than divalent cations, because the ratio of replacement is increased from 2:5 (5.17) to 3:5 (5.21).

6. Strontium Barium Niobate

The ferroelectric oxide SBN ($(Sr_{1-x}Ba_x)Nb_2O_6$) exists as a solid solution with compositions corresponding to $0.25 \leq x \leq 0.75$. The material shows pronounced photorefractive properties due to the large electrooptic and dielectric material parameters (see [304] for a compilation of physical properties). The corresponding figure of merit is better than for $BaTiO_3$ and $KNbO_3$ [2], but depends on x.

The crystal stucture of SBN corresponds to the tetragonal tungsten bronze structure. It may be characterized (viewing along the c-axis) as being built up of two perovskite-structured slabs linked with two additional sequences of corner-sharing NbO_6 octahedra. The realization of this structure requires significant distortions of the octahedra. The ferroelectric space group is $P4bm$, of which the unit cell contains five formula units. Nb ions have octahedral coordination spheres corresponding to the 2b and 8d lattice positions. The complete corner-sharing NbO_6 framework gives rise to three types of channels along the c-axis with triangular, square and pentagonal shapes, which contain the ninefold lattice site (denoted C subsequently), the twelvefold site 2a and the fifteenfold lattice site 4c, respectively. The 4c lattice sites are partially occupied by Ba and Sr, but Sr is also found on the 2a sites. Interstitial sites refer to vacant 2a and 4c sites as well as the ninefold coordinated lattice sites C. The C positions remain completely unoccupied. There is also significant disorder on the oxygen sublattice.

The ferroelectric Curie temperature ranges between about 50 and 200 °C and correlates directly with increasing x. For $T > T_c$ all crystal ions are located within equidistant mirror planes perpendicular to c. Below T_c the metal ions move off their respective mirror planes giving rise to the spontaneous polarization $\boldsymbol{P}_s$. The displacements are largest for niobium ions [6].

So far, no electronic structure calculations have been reported for this complex oxide. Recently, Baetzold [305] reported shell model-based simulations of the material's defect chemistry employing $x = 0.4$. The most important results of these investigations will now be reviewed.

The potential parameters have been transferred from SrO, BaO and $LiNbO_3$. In order to account for the random lattice site occupations, particularly of the Ba and Sr cations, a random number routine has been employed. However, the simulations reveal that random distributions of crystal ions affect the lattice energy of the completely relaxed perfect crystals only marginally. The energy variations amount to a few hundreths of an eV per formula unit $(Sr_{1-x}Ba_x)Nb_2O_6$ and may thus be safely neglected. The corresponding energy differences are considerably larger (a few eV per $(Sr_{1-x}Ba_x)Nb_2O_6$) prior to the necessary lattice equilibrations.

The most important results concerning defect formations can be summarized as follows: the intrinsic defect structure of SBN is determined by randomness of Sr^{2+} and Ba^{2+} cations. For these cations Frenkel disorder and the formation of $Ba_{Sr}^{\times}$ and $Sr_{Ba}^{\times}$ antisite defects are likely to occur. Whereas the Sr/Ba cation interchange reaction

$$\mathrm{Sr_{Sr}^{\times}} + \mathrm{Ba_{Ba}^{\times}} \longrightarrow \mathrm{Sr_{Ba}^{\times}} + \mathrm{Ba_{Sr}^{\times}} \tag{6.1}$$

needs 0.6 eV per defect, the formation of Frenkel defects requires only 0.3 eV per defect. These small energy values corrrelate directly with the corresponding results of perfect lattice simulations quoted above. On the other hand, Schottky-type defects are not favoured to occur. Moreover, due to the large Nb vacancy formation energy, disorder involving these cations seems to be unlikely.

Impurities with electronic states in the band gap are indispensable for the photorefractive effect. As in the other materials discussed in this volume, the impurity charge states and incorporation sites are decisive with respect to donor/acceptor behaviour. The investigations of Baetzold are restricted to the donor levels of defects, which respond to electron transport in the conduction band.

In most instances trivalent and tetravalent impurity cations prefer an incorporation by means of self-compensation. For the dominant cerium and iron impurities this results in Nb and Ba/Sr site incorporations. In the case of Ce^{4+} self-compensation reads as:

$$\mathrm{CeO_2} +' \mathrm{SBN}' \longrightarrow \frac{1}{3}Ce_{Sr}^{\bullet\bullet} + \frac{2}{3}Ce_{Nb}^{l} + \frac{1}{3}\mathrm{Nb_2O_5} + \frac{1}{3}\mathrm{SrO} \tag{6.2}$$

$$\mathrm{CeO_2} +' \mathrm{SBN}' \longrightarrow \frac{1}{3}Ce_{Ba}^{\bullet\bullet} + \frac{2}{3}Ce_{Nb}^{l} + \frac{1}{3}\mathrm{Nb_2O_5} + \frac{1}{3}\mathrm{BaO}\,. \tag{6.3}$$

Similar reactions may be formulated for trivalent cations. Most competitive to self-compensation is the formation of charge-compensating Ba or Sr vacancies, e.g. for Ce^{4+}:

$$\mathrm{CeO_2} +' \mathrm{SBN}' \longrightarrow Ce_{Sr}^{\bullet\bullet} + V_{Sr}^{ll} + 2SrO \tag{6.4}$$

$$\mathrm{CeO_2} +' \mathrm{SBN}' \longrightarrow Ce_{Ba}^{\bullet\bullet} + V_{Ba}^{ll} + 2BaO\,. \tag{6.5}$$

Typically self-compensation and vacancy-assisted solution differ in energy (per molecule of the impurity oxide) by a few tenths of an eV. In some cases the latter incorporation mode is even preferred. For example, trivalent manganese is found at the Ba/Sr sites according to the reaction:

$$\mathrm{Mn_2O_3} +' \mathrm{SBN}' \longrightarrow 2\mathrm{Mn_{Ba}^{\bullet}} + \mathrm{V_{Ba}^{ll}} + 3\mathrm{BaO}\,. \tag{6.6}$$

Whereas a minority of the impurity cations can be incorporated at interstitial 2a and 4c sites, there are no theoretical indications of impurity incorporations at the ninefold coordinated lattice site C.

Divalent cations consistently prefer an incorporation at Ba/Sr sites. There is no requirement for charge compensations.

Donor states of impurity cations, such as Ce^{3+}, Fe^{2+} and Mn^{2+}, are stable on the Ba/Sr sites, but would be ionized in the case of Nb site incorporation. If electronic transport takes place via the conduction band, the stable donor states will provide the relevant defect levels for the photorefractive effect to be operative. The donor levels have been estimated by combining defect formation energies and ionization potentials of free ions according to

$$\mathrm{M}^{n+} \longrightarrow \mathrm{M}^{(n+1)+} + \mathrm{Nb}_{\mathrm{Nb}}^{4+}, \tag{6.7}$$

where $\mathrm{Nb}_{\mathrm{Nb}}^{4+}$ ions represent conduction band states which have been simulated assuming small polarons to be favourable. It seems noticeable that iron and cerium may possess several energy levels in the band gap. This result would be related to the existence of different Ba and Sr sites in the material.

Due to the use of free-ion ionization potentials the results should not be interpreted quantitatively, but the results might provide helpful qualitative predictions. However, significant uncertainties may be due to the neglect of covalency differences and crystal field splitting contributions, which are particularly important in the case of 3d transition metal impurities. Quantitative estimates, on the other hand, would demand the employment of embedded cluster calculations including lattice relaxation.

7. Yttrium Iron Garnet

Garnet crystals play an important role in many technological devices. For example, yttrium aluminium garnet (YAG) doped with Nd^{3+} ions is one of the most commonly used laser materials. Further examples include magnetic garnets, of which the ferrimagnetic yttrium iron garnet (YIG) represents the most prominent member. YIG devices are used extensively in various microwave applications [3].

Recent developments concern the construction of magnetooptical device components for their use in integrated optics. The corresponding techniques are based on thin YIG films deposited on substrate garnet materials [3, 4, 5]. Similar to the electrooptic materials the basic physical crystal properties of YIG (e.g. optical absorption, Faraday rotation and photomagnetic effects) depend to a considerable extent on its defect structure. For example, Fe^{2+} and Fe^{4+} ions as well as Bi^{3+} dopants strongly modify the magnetooptical crystal properties. Similar influences can be ascribed to Pb^{2+} impurities which are always present in YIG as part of the usual crystal growth techniques [3]. Garnet crystals are well suited to accommodating many different cation types with various valencies and with concentrations in the percentage range [3]. The optimization of technical properties of the material should be facilitated by controlled impurity incorporation – a procedure which may be described as "molecular engineering".

Because there are no electronic structure calculations for complex garnet crystals, the present chapter is restricted to providing a detailed discussion of shell model-based potential simulations. The emphasis is placed on the defect chemistry of YIG. In spite of the extensive experimental work in this field, there is still a need for a definite and generally accepted defect model, which covers the basic questions referring to the dominant intrinsic defects as well as the incorporation sites and modes of charge compensation of impurity ions. Certainly, the corresponding knowledge is the first important step towards a successful molecular engineering.

7.1 Potential Models

In the present simulation study on YIG we consider two sets of shell model potential parameters (Table 7.1).

Table 7.1. Shell model parameters according to parameter sets 1 and 2

Interaction		A (eV)	ϱ (Å)	C (eV Å^6)
O^{2-}... O^{2-}	1	22764.0	0.149	27. 8
	2	22764.0	0.149	87.5
Y^{3+}... O^{2-}	1	1345.1	0.3491	0.0
	2	1388.0	0.3561	0.0
$Fe^{3+}_{(a)}$... O^{2-}	1	1102.4	0.3299	0.0
	2	993.9	0.3400	0.0
$Fe^{3+}_{(d)}$... O^{2-}	1	1102.4	0.3299	0.0
	2	852.3	0.3490	0.0

Ion		Y(| e |)	k (eV Å^{-2})
O^{2-}	1	−2.811	103.07
	2	−3.148	43.31
Y^{3+}	1	–	–
	2	–	–
$Fe^{3+}_{(a)}$	1	4.97	304.7
	2	5.30	408.0
$Fe^{3+}_{(d)}$	1	4.97	304.7
	2	4.65	205.0

The first (1) was obtained by transferring empirical shell model parameters from Y_2O_3 and α-Fe_2O_3 [122] to YIG, while the second (2) was derived by empirical fitting to the properties of YIG. In order to keep the number of degrees of freedom per unit cell manageable, the yttrium ions have been defined to be unpolarizable in both parameter sets, both of which reproduce the observed crystal structure to sufficient accuracy (neglecting slight distortions arising from the ferrimagnetic order in YIG) [3]. Deviations with respect to ion positions are less than 0.05 Å. It is noted that magnetic interactions mediated by superexchange cannot be accounted for within classical shell model simulations.

Garnet crystals possess a cubic body-centred Bravais lattice belonging to the space group Ia 3d (O_h^{10}). The basis consists of four formula units of YIG ($\hat{=}$ $Y_3Fe_5O_{12}$), where the Y^{3+} ions occupy the 24c dodecahedrally coordinated sites and Fe^{3+} ions both the octahedral 16a and the tetrahedral 24d sites. More accurately one should write $(Y_3)_{24c}(Fe_2)_{16a}(Fe_3)_{24d}O_{12}$. Figure 7.1 (reproduced from [3]) shows the spatial connection of the various oxygen polyhedra existing in YIG.

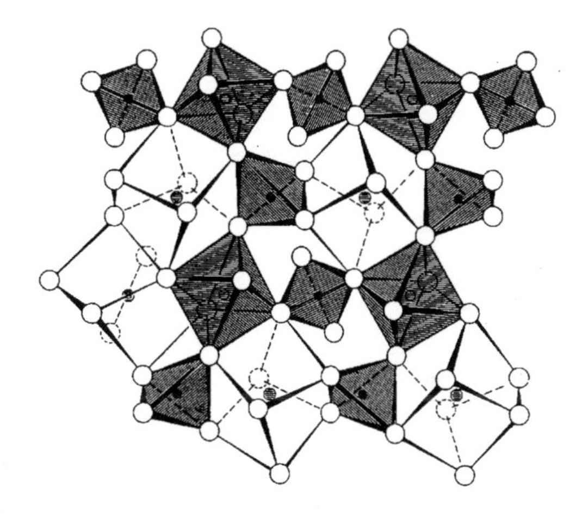

Fig. 7.1. Spatial connection of oxygen octahedra in garnets. The framework of alternating tetrahedra and octahedra (shaded) and of eightfold dodecahedra is shown. Large open circles represent oxygen anions; small circles represent cations

By inspection of Table 7.1 we observe that in set 2 tetrahedrally and octahedrally coordinated Fe^{3+} ions have been assigned different interatomic potential parameters, which in principle is reasonable due to possible changes of cation properties (e.g. ion size) as a function of coordination number. In YIG, however, such coordination dependent differences turn out to be small.

Table 7.2 summarizes calculated and experimentally determined macroscopic constants.

It is seen that only set 2 can accurately reproduce the dielectric behaviour of YIG, which is of considerable importance in predicting reliable defect energies. All defect energies reported below refer to the parameter set 2, except when indicated to the contrary.

Figure 7.2 displays a typical region I used in the subsequent Mott–Littleton defect calculations (see also Sect. 2.2).

Table 7.2. Comparison of calculated and measured macroscopic constants

Macroscopic constant	Calculated		Experiment [3, 306]
	set 1	set 2	
C_{11} (GPa)	326.0	273.4	268.0 – 270.1
C_{12} (GPa)	111.5	129.8	110.6 – 110.9
C_{44} (GPa)	101.2	68.7	76.6 – 77.5
ε_0	10.3	17.5	$\simeq 17$
ε_∞	2.11	5.3	$\simeq 5$

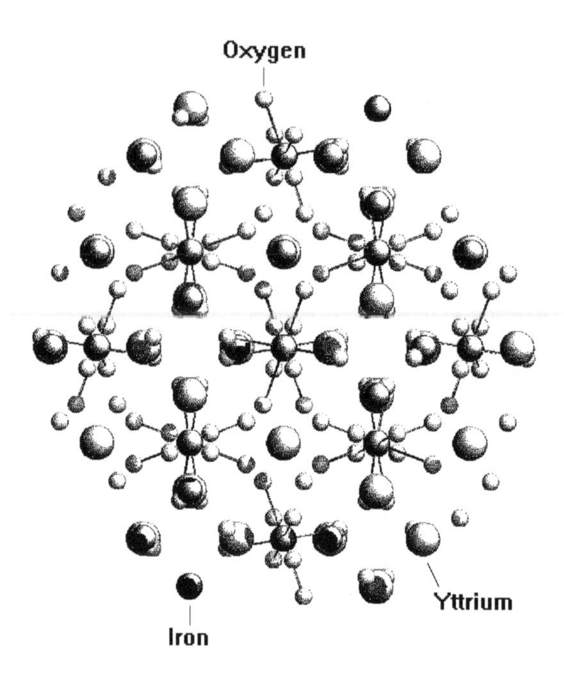

Fig. 7.2. A typical region I used in the defect simulations in YIG. This crystal region which immediately surrounds an octahedrally coordinated iron site defect is explicitly relaxed according to the underlying potential model. The direction of viewing is perpendicular to a (100) crystal plane

7.2 Chemical Stability of YIG

Having demonstrated the success of the potential models in describing structural properties of YIG we continue with consideration of the extent to which they are compatible with the observed stoichiometry variations and phase relations of the material.

Under most practical conditions the phase diagram appropriate to YIG can be described by the binary Y_2O_3–Fe_2O_3 system. Deviations from this description will only occur if the system is exposed to reducing or oxidizing atmospheres [3, 307]. In this latter case the analysis must include the relevant defect redox reactions.

At the present stage it is sufficient to take advantage of the "pseudo-binary" system and thus to consider reactions of the following type:

$$2Y_3Fe_5O_{12} \longrightarrow 3Y_2O_3 + 5Fe_2O_3 \tag{7.1}$$

or alternatively

$$Y_3Fe_5O_{12} \longrightarrow Fe_2O_3 + 3YFeO_3\,. \tag{7.2}$$

The energies of reaction can be estimated by combining calculated lattice energies (per formula unit) using the YIG potential parameters (Table 7.1). In

(7.2) we have included the perovskite structured yttrium orthoferrite $YFeO_3$, which is readily formed on the yttrium-rich side of the phase diagram. It is also noted that Y_2O_3, Fe_2O_3 and $YFeO_3$ are fairly well modelled using the YIG parameters. For example, the calculated lattice constant of $YFeO_3$, $a =$ 3.863 Å, is only 0.016 Å larger than the experimental value.

As a result one obtains (per formula unit YIG) +1.56 eV and +2.91 eV for reactions (7.1) and (7.2), respectively. These values show that the stoichiometric formation of YIG is favourable. The same qualitative conclusion is reached if parameter set 1 is used instead of 2.

We conclude this section with some preliminary considerations on non-stoichiometry, to which we return later when discussing defect calculations. We consider both Y_2O_3 and Fe_2O_3 excess in YIG crystals by modelling $Y_3(Fe_4Y)O_{12}$ and $(Y_2Fe)Fe_5O_{12}$ respectively. Thus in the first case 1/5 of the Fe in the garnet is replaced by Y, and in the second 1/3 of the Y is replaced by Fe. This procedure, despite the fact that the compositions show greater Y and Fe excess than observed experimentally, nevertheless allow us to probe the energetics associated with non-stoichiometry. Thus calculations are performed using the parameter sets in Table 7.1 for the following reactions:

$$Y_2O_3 + 3Fe_2O_3 \longrightarrow (Y_2Fe)Fe_5O_{12} \tag{7.3}$$

for which the calculated energy is 1.33 eV and

$$2Y_2O_3 + 2Fe_2O_3 \longrightarrow Y_3(Fe_4Y)O_{12} \tag{7.4}$$

where −0.77 eV is calculated for the energy of reaction.

Whereas Fe_2O_3 excess turns out to be slightly endothermic, Y_2O_3 excess is seen to be exothermic. At first sight this result seems to contradict the observed phase diagram, from which a small possible Fe_2O_3 excess can be inferred but no Y_2O_3 excess. In order to resolve this contradiction one can suggest that the formation of $YFeO_3$ might be preferred over Y_2O_3-rich YIG. Indeed, by calculation of appropriate lattice energies, and taking into account that one unit of $Y_3(Fe_4Y)O_{12}$ corresponds to four units of $YFeO_3$, this interpretation turns out to be reasonable, by reference to the following lattice energies:

$$\begin{aligned} E_{\text{Latt}}(YFeO_3) &= -141.1\text{eV}\,(-142.5\,\text{eV}) \\ \frac{1}{4}\,E_{\text{Latt}}(Y_3(Fe_4Y)O_{12}) &= -141.7\text{eV}\,(-141.9\,\text{eV}). \end{aligned}$$

(where energies in brackets refer to parameter set 1). We should approach these results with some caution, owing to the uncertainties introduced by transferring YIG parameters to different materials as well as by neglecting any thermal effects. Nevertheless the qualitative conclusions are reasonably clear: yttrium excess may be accommodated by formation of the perovskite $YFeO_3$ rather than non-stoichiometric phases, as there is very little difference in the lattice energies of the two phases.

7.3 Intrinsic Defect Structures

We now discuss point defect formation mechanisms involving yttrium, iron or oxygen ion species only. Impurity point defects will be considered in Sect. 7.5.

Table 7.3 lists the basic defect formation energies, from which it is straightforward to consider the energetics of Schottky- and Frenkel-type defect formation.

Table 7.3. Basic defect formation energies using parameter set 2

Defect	Defect formation energy (eV)
$\mathrm{V}_{\mathrm{Fe(a)}}^{3l}$	51.49
$\mathrm{V}_{\mathrm{Fe(d)}}^{3l}$	50.88
$\mathrm{V}_{\mathrm{O}}^{\bullet\bullet}$	21.43
$\mathrm{V}_{\mathrm{Y}}^{3l}$	42.96
$\mathrm{Fe}_{\mathrm{Y}}^{\times}$	−8.03
$\mathrm{Y}_{\mathrm{Fe(a)}}^{\times}$	9.63
$\mathrm{Y}_{\mathrm{Fe(d)}}^{\times}$	10.57
$\mathrm{O}_{\mathrm{I}}^{ll}$	−14.32
$\mathrm{Fe}_{\mathrm{I}}^{3\bullet}$	−39.02
$\mathrm{Y}_{\mathrm{I}}^{3\bullet}$	−26.05

In all subsequent reactions the defect notation of Kröger and Vink [205] will be used; the symbol 'YIG' denotes the bulk crystal.

Schottky Disorder:

$$'\mathrm{YIG}' \longrightarrow \mathrm{Fe_2O_3} + 2\mathrm{V}_{\mathrm{Fe}}^{3l} + 3\mathrm{V}_{\mathrm{O}}^{\bullet\bullet} \tag{7.5}$$

$$'\mathrm{YIG}' \longrightarrow \mathrm{Y_2O_3} + 2\mathrm{V}_{\mathrm{Y}}^{3l} + 3\mathrm{V}_{\mathrm{O}}^{\bullet\bullet} \tag{7.6}$$

$$'\mathrm{YIG}' \longrightarrow \mathrm{Y_3Fe_5O_{12}} + 3\mathrm{V}_{\mathrm{Y}}^{3l} + 5\mathrm{V}_{\mathrm{Fe}}^{3l} + 12\mathrm{V}_{\mathrm{O}}^{\bullet\bullet} \tag{7.7}$$

$$'\mathrm{YIG}' \longrightarrow \mathrm{YFeO_3} + \mathrm{V}_{\mathrm{Y}}^{3l} + \mathrm{V}_{\mathrm{Fe}}^{3l} + 3\mathrm{V}_{\mathrm{O}}^{\bullet\bullet} \tag{7.8}$$

(where only reaction (7.7) represents true Schottky disorder, as the others result in a change in the chemical composition of the material)

Frenkel Disorder:

$$'\mathrm{YIG}' \longrightarrow \mathrm{V}_{\mathrm{Fe}}^{3l} + \mathrm{Fe}_{\mathrm{I}}^{3\bullet} \tag{7.9}$$

$$'\mathrm{YIG}' \longrightarrow \mathrm{V}_{\mathrm{Y}}^{3l} + \mathrm{Y}_{\mathrm{I}}^{3\bullet} \tag{7.10}$$

$$'\mathrm{YIG}' \longrightarrow \mathrm{V}_{\mathrm{o}}^{\bullet\bullet} + \mathrm{O}_{\mathrm{I}}^{ll}\,. \tag{7.11}$$

The respective energies are summarized in Table 7.4.

Table 7.4. Reaction energies (per defect) for Schottky-like and Frenkel disorder reactions

Disorder reaction	Reaction energy per defect (eV)
Schottky-like	
(7.5)	3.11
(7.6)	3.54
(7.7)	3.19
(7.8)	3.41
Frenkel	
(7.9)	5.93
(7.10)	8.45
(7.11)	3.55

The creation of iron vacancies is assumed to involve tetrahedral iron sites only, because corresponding defect energies are smaller compared with octahedral sites (see Table 7.3). Even the most favourable defect reactions in Table 7.4 need for their creation large formation energies per defect of about 3.1-3.5 eV. In accordance with previous investigations [122, 124] it can be concluded that Schottky- and Frenkel-type disorder are insignificant on energetical grounds.

A different type of intrinsic disorder may be introduced by interchanging Y^{3+} and Fe^{3+} cations according to the solid state reaction

$$Y_Y^\times + Fe_{(a)}^\times \longrightarrow Y_{(a)}^\times + Fe_Y^\times, \tag{7.12}$$

for which is calculated a reaction energy of 0.8 eV per defect. The shortened subscript (a) denotes the octahedral iron site. The comparatively small reaction energy suggests that this type of disorder should exist in most YIG crystals. Experimental evidence in favour of antisite defects occurring as in (7.12) has been reported recently by Wagner et al. [308]: The satellite signals in the NMR spectra of ^{57}Fe in nominally pure YIG could be partly assigned to existing antisite defects. The cation interchange specified above does not alter the magnetic moment per formula unit, because the 24c and 16a sublattices have parallel spin alignment [3]. The defect chemical results discussed so far refer to YIG crystals which are perfectly grown otherwise, but as most YIG crystals are known to be affected by considerable impurity concentrations (e.g. Bi^{3+} ions), corresponding influences should be included. Importantly, the cation interchange energy is reduced in all crystals which are doped with large cations. This is verified, for example, by simulating the cation interchange in $Y_3(Fe_4Y)O_{12}$ (0.54 eV per defect) and in $(Y_{\frac{9}{4}}Bi_{\frac{3}{4}})Fe_5O_{12}$ (0.70 eV per defect). For modelling of the latter compound a Bi...O short-range interaction potential has been transferred from preliminary simulation studies on $BaBiO_3$ [309].

A further source for antisite defects provides the accommodation of non-stoichiometry. Generally, the obvious manifestations of non-stoichiometry, i.e. Fe_2O_3 or Y_2O_3 excess, may be compensated by antisite defects, interstitial ions or vacancies:

$$\frac{1}{2}Y_2O_3 +' YIG' \longrightarrow Y^{\times}_{(a)} + \frac{1}{2}Fe_2O_3 \tag{7.13}$$

$$Y_2O_3 +' YIG' \longrightarrow 2Y^{3\bullet}_{I} + 3O^{ll}_{I} \tag{7.14}$$

$$Y_2O_3 +' YIG' \longrightarrow 2YFeO_3 + 2V^{3l}_{Fe} + 3V^{\bullet\bullet}_{O} \tag{7.15}$$

$$\frac{1}{2}Fe_2O_3 +' YIG' \longrightarrow Fe^{\times}_{Y} + \frac{1}{2}Y_2O_3 \tag{7.16}$$

$$Fe_2O_3 +' YIG' \longrightarrow 2Fe^{3\bullet}_{I} + 3O^{ll}_{I} \tag{7.17}$$

$$Fe_2O_3 +' YIG' \longrightarrow 2YFeO_3 + 2V^{3l}_{Y} + 3V^{\bullet\bullet}_{O}. \tag{7.18}$$

Inspection of Table 7.5 shows that the formation of interstitial ions and vacancies is energetically unfavourable. It should be further noted that in agreement with the previous calculations of chemical stability (Sect. 7.2), excess of Y_2O_3 is preferred to surplus Fe_2O_3. However, as we have already discussed in Sect. 7.2 the formation (or phase separation) of yttrium orthoferrite is likely to reduce the occurrence of excess Y_2O_3 in YIG crystals. Fe_2O_3 excess is

Table 7.5. Energetics of Y_2O_3 and Fe_2O_3 excess in YIG

Reaction	Reaction energy (eV) per molecule A_2O_3
(7.13)	1.24
(7.14)	37.44
(7.15)	16.43
(7.16)	1.96
(7.17)	29.52
(7.18)	18.61

well known to occur in otherwise pure YIG [310, 311]. In contrast, evidence for excess Y_2O_3 has only been found in YIG doped with one aluminium ion per formula unit [312]. Finally, the formation of $Fe^{\times}_{Y}$ antisites resulting from surplus Fe_2O_3 should induce a decrease in the lattice constant due to ion-size differences. Corresponding support is given by various experimental investigations [310, 313]. For example, Hansen et al. [313] found strong evidence for Fe_Y antisite defects by studying the lattice constant of YIG as a function of Fe_2O_3 non-stoichiometry. The lattice constant is significantly smaller in non-stoichiometric crystals than in stoichiometric samples. On the basis of ion-size arguments a concentration of ~0.01 Fe_Y antisites per formula unit YIG has been reported. This value is consistent with the known Fe_2O_3 non-stoichiometry. However, the actual antisite concentration could be significantly larger if interchange reactions are considered as discussed in (7.12). These reactions are even more favourable than processes related to non-stoichiometry. Since both Fe_Y and $Y_{(a)}$ antisites are the products of such interchange reactions, there are probably no measurable effects upon the lattice constant. In any case, the antisite concentration is sufficiently large in order to influence other defect formation mechanisms.

7.4 Intrinsic Electronic Properties

In this section we remark on reduction and oxidation treatments, electrical conductivity and optical absorption mechanisms in YIG crystals, within the obvious limitations imposed by a treatment based on effective potentials. Redox processes induce the formation of additional $Fe^{2+/4+}$ ion species which are modelled by appropriately changing the ionic charge states with respect to the intrinsic 3+ state. As the modification of short-range potential parameters, at the same time, is expected to be small, these parameters are taken to be unaltered.

YIG crystals annealed in atmospheres with low oxygen partial pressure contain additional electrons which lead to the formation of ferrous ions, Fe^{2+}. Even under growth conditions Fe^{2+} ions have been observed in YIG [314]. It has been suggested that these extra charges may be compensated either by oxygen vacancies or by interstitial ions [310]. Shell model calculations confirm that oxygen vacancies are more likely to occur:

$$'\mathrm{YIG}' \longrightarrow \frac{3}{2}\mathrm{O}_2(\mathrm{g}) + 3V_{\mathrm{O}}^{\bullet\bullet} + 6\mathrm{Fe}^{\prime} \tag{7.19}$$

$$'\mathrm{YIG}' \longrightarrow \frac{15}{4}\mathrm{O}_2(\mathrm{g}) + 5\mathrm{Fe}_{\mathrm{I}}^{3\bullet} + 15\mathrm{Fe}^{\prime} - \mathrm{Y}_3\mathrm{Fe}_5\mathrm{O}_{12} + \frac{3}{2}\mathrm{Y}_2\mathrm{O}_3\,, \tag{7.20}$$

where (7.20) represents the most favourable reduction mechanism involving interstitial ions. It should be noted that the interstitial type reaction requires one formula unit of YIG to redissolve from the surface into the bulk lattice. However, this process is energetically costly: (7.19) is by about 6 eV per O_2 molecule more favourable than (7.20).

The absolute reduction energies depend on the precise formation energy of Fe^{2+} ions. These may be estimated by combining shell model defect energies with the appropriate free-ion ionization energies [315]. In perfectly grown YIG octahedrally coordinated Fe^{2+} ions are by 0.6 eV more favourable than tetrahedrally coordinated Fe^{2+} ions. Possible corrections due to differences between the octahedral and tetrahedral crystal field splitting contributions are small and range between 0.0 and 0.2 eV in favour of the tetrahedral site, where the precise correction depends on the employed cubic crystal field splitting energies usually denoted by 10Dq. If $Fe_Y^{\times}$ antisite defects are present (e.g. resulting from non-stoichiometry or from cation interchange discussed above) these defects become the favoured "traps" for additional electrons. The existence of antisite defects lowers the reduction energy by further 4.6 eV per O_2, and the reaction energy of (7.19) becomes 11.9 eV per O_2. Despite the considerable amount (see also below) this value is by about 25 eV per O_2 smaller than the reduction energy calculated for YAG [316], which is in qualitative agreement with the observation that YIG can be to some extent reduced, but not YAG [316]. Table 7.6 summarizes binding energies between oxygen vacancies and trapped electrons. The binding energy is defined as the reaction energy corresponding to $Fe^{l} + V_O^{\bullet\bullet} \rightarrow (Fe^{l}V_O^{\bullet\bullet})$ (complex). The inclusion of binding energies reduces the calculated reduction energies even further. The last defect complex in Table 7.6 is, however, assumed to be defect chemically insignificant because of the comparatively low concentrations of the constituent defect species. All reduction mechanisms discussed so far can be used to describe the observed thermal decomposition of YIG which takes place under strong reducing conditions [317]. For example, (7.19) may be reformulated as follows:

$$\begin{aligned} 6Y_3Fe_5O_{12} &\longrightarrow O_2 + 2V_O^{\bullet\bullet} + 4Fe^{l} + 26Fe_{Fe}^{\times} + 18\,Y_Y^{\times} + 70\,O_O^{\times} \\ &\longrightarrow O_2 + 18YFeO_3 + 4(Fe^{l} + 2Fe_{Fe}^{\times} + 4O_O^{\times}) \\ &\longrightarrow O_2 + 18YFeO_3 + 4Fe_3O_4\,. \end{aligned} \tag{7.21}$$

Table 7.6. Binding energies (eV) of electrons at oxygen vacancies

Defect complex	Binding energy (eV)
$(V_O^{\bullet\bullet}\ Fe^{l}_{(a)})$	−0.38
$(V_O^{\bullet\bullet}\ Fe^{l}_{(d)})$	−0.56
$(V_O^{\bullet\bullet}\ Fe^{l}_{Y})$	−0.85

We now consider the oxidation of YIG in oxygen-rich atmospheres. The following reaction refers to perfectly grown YIG crystals:

$$O_2(g) +'\,YIG' \longrightarrow 2O_I^{ll} + 4Fe^{\bullet}_{(d)}\,, \tag{7.22}$$

for which one calculates an energy of 17.6 eV per O_2 molecule which is prohibitively high. However, in most YIG crystals oxygen vacancies will be present either as frozen-in from high-temperature reducing growth conditions or as charge compensators for divalent impurities. Later oxidation treatments may then be viewed as simply filling oxygen vacancies:

$$4\mathrm{Fe}_{(\mathrm{d})}^{\times} + \mathrm{O}_2(\mathrm{g}) + 2\mathrm{V}_{\mathrm{O}}^{\bullet\bullet} \longrightarrow 2\mathrm{O}_{\mathrm{O}}^{\times} + 4\mathrm{Fe}_{(\mathrm{d})}^{\bullet} \,. \tag{7.23}$$

The oxidation energy in this latter case is calculated to be 3.3 eV per O_2. In conclusion, the oxidation of YIG will have a reasonably small reaction energy only if there exists an oxygen vacancy population prior to oxidation. Note that holes at tetrahedral iron sites are more favourable than holes localized at octahedral iron sites (by 4.2 eV) as well as at oxygen sites (by 1.6 eV).

It must be emphasized that the calculated redox reaction energies are very sensitive to the accuracy with which electronic energy terms can be accounted for in shell model simulations (note the amplifying factor of four per O_2 molecule occuring in (7.19)–(7.23)). Thus, using free-ion ionization potentials for iron would probably overestimate redox energies by a few eV. The discussion in Sect. 3.2 confirms that the deviations of in-crystal ionization potentials from their free-ion counterparts essentially result from covalent charge transfer and crystal field splitting terms. It is possible that, due to filled antibonding d states of iron, the covalency contributions are smaller than the crystal field splitting terms, which would enhance the electron affinity of YIG in comparison with calculated affinities based on free-ion ionization potentials. Thus, the embedding lattice-induced cationic level shifts might be directed oppositely to the case of the oxide perovskites (see Sect. 3.2). The qualitative importance of our present estimates, however, in predicting the favourable redox processes is obvious, since the possible shortcomings regarding the electronic structure have been noted consistently throughout the present work. Moreover, in order to determine the precise driving forces of redox reactions, i.e. the Gibbs free energy $\Delta G = \Delta H - T\Delta S$, it is necessary to include relevant entropy terms referring to the formation (reduction) or consumption (oxidation) of oxygen gas (e.g. using standard entropy data for oxygen one finds $T\Delta S \simeq 2.2$ eV at $\mathrm{T} \simeq 600$ °C). Thus, $\Delta G < \Delta H$ is obtained in the case of reduction, but $\Delta G > \Delta H$ for oxidation. The inclusion of entropy terms, however, would not influence the qualitative predictions of the most favourable reduction and oxidation mechanisms.

Macroscopic crystal properties such as electrical conductivity and optical absorption are significantly influenced by existing Fe^{2+} and Fe^{4+} ions. These species may be investigated by means of shell model calculations, which have the advantage that lattice relaxation effects are properly taken into account. Quantum mechanical terms, on the other hand, can only be considered in a very approximate way by adding appropriate free-ion ionization energies and (where necessary) crystal field stabilization energies. A shell model treatment of electron or hole species necessarily implies a small-polaron model. The stability condition favouring small polarons is given by

$$|E_{\mathrm{B}}| > \frac{\triangle}{2} \,, \tag{7.24}$$

where E_{B} means the small-polaron binding energy and $\triangle$ the appropriate band width. The right-hand side of (7.24) represents the delocalization energy gain within a band model, which may be considered to be competitive with small-polaron formation. In the case of electrons the band width $\triangle = 0.6$ eV is derived from the octahedral t_{2g} (Fe^{2+}) states [318]. For the small-polaron binding energy, which is determined from the lattice relaxation around an octahedrally coordinated Fe^{2+}, $|E_B|$=1.3 eV is obtained. This value is the difference between the energy calculated with full lattice relaxation and that obtained when only shells are allowed to relax and the cores (representing the nuclei) are frozen. If we further include the crystal field splitting energy and its dependence upon nearest neighbour distances (10Dq $\propto (r_{ML})^{-5}$) $|E_B|$ is finally reduced to about 1.18 eV assuming 10Dq $\simeq$ 1 eV for the optical (unrelaxed) case and taking into account the calculated outward displacement of the oxygen ligands ($\simeq$ 0.15 Å) in the fully relaxed (thermal) case. Thus, on the basis of (7.24) we find that the small electron polaron is stable in pure YIG. We should emphasize that the reliability of the calculated binding energy E_B depends to some extent on the ability of our shell model parameter set (2) to model the dielectric properties of YIG with sufficient accuracy.

Consequently, polaron hopping seems to represent the dominant electronic conductivity mechanism in YIG. However, it is difficult to prove this conjecture experimentally as well as theoretically. First, there is a considerable scatter in measured conductivity data resulting from differing experimental conditions regarding the precise chemical composition and type of the YIG samples, the growth conditions and the experimental techniques employed, etc. The conductivity activation energy is found to vary between $\simeq$0.3 eV and $\simeq$1.8 eV [3]. Very careful conductivity measurements have been performed in the case of Si-doped YIG single crystals [318]. The conductivity activation energy (0.3 eV) has been interpreted as the binding energy of electrons at Si^{4+} dopant ions with no appreciable contributions from mobility activation (< 0.1 eV), which led to the assumption of band-like electrons. Next, it is recalled that small-polaron theories provide different regimes of possible polaron hopping depending on temperature, phonon energies, adiabaticity and on jump correlations [319, 320, 321]. As a result, in some cases the drift and Hall mobilities need not manifest a clear thermally activated behaviour. Thus, the experimentally observed absence of mobility activation, as was claimed in Si-doped YIG, cannot definitely prove the non-existence of small polarons.

Assuming the validity of the non-adiabatic hopping regime in the high-temperature limit [319] allows us to calculate an upper bound for the polaron hopping activation energy in (pure) YIG. One calculates $E_H = 0.6$ eV following the approach of Norgett and Stoneham [322], in which the saddle-point energy is obtained by first distributing the electron charge equally over the two neighbouring cation sites and allowing the lattice to relax to equilibrium. This configuration is then frozen and the energy recalculated with the electron localized on one of the sites. This value is in good agreement with the prediction $E_H = \frac{1}{2}|E_B|$ from the Holstein "molecular crystal model" [323]. It is, however, too large compared with the results for Si-doped YIG [318], indicating that the non-adiabatic high-temperature regime is probably not applicable in this case.

Beyond the "mobility problem" we have, as in the case of band-like electrons, to consider possible electron traps leading to an additional contribution to the activation energy related to the charge carrier concentration. Intrinsic trapping centres are provided by oxygen vacancies $V_O^{\bullet\bullet}$ (mainly resulting from prior reduction treatments or possibly from charge compensation of divalent impurities) and by $Fe_Y^{\times}$ antisite defects. Thus, the reactions

$$(V_O^{\bullet\bullet}\ Fe_{(a)}^{l}) \longrightarrow V_O^{\bullet\bullet} + Fe_{(a)}^{l} \qquad (7.25)$$

$$Fe_{(a)}^{\times} + Fe_Y^{l} \longrightarrow Fe_{(a)}^{l} + Fe_Y^{\times} \qquad (7.26)$$

define corresponding electronic charge carrier activation modes. The reaction energies are calculated to be 0.38 eV for (7.26) (see Table 7.6) and 1.16 eV for (7.26). The reaction energy for (7.26) involving the predicted Fe_Y antisite defects has been calculated by neglecting corrections due to crystal field splittings and other electronic energy contributions which are not accounted for within the shell model. If, for example, the crystal field splitting of Fe_Y antisites is to some extent smaller than the corresponding term of $Fe_{(a)}$ ions, the reaction energy for (7.26) may be substantially lower than our calculated value.

Extrinsic trapping centres are given by higher valent impurity ions as Si^{4+}. The conductivity activation energy of 0.3 eV for Si–YIG [318] has been interpreted as the binding energy of $(Si_{(d)}^{\bullet}\ Fe_{(a)}^{l})$ defect complexes. The calculated binding energy of 0.28 eV obviously provides some support for this interpretation.

In the case of holes as dominant charge carriers the existence of small-hole polarons remains unclear, because the appropriate valence band width is not known exactly. It has been estimated as $\simeq$4 eV [324]. However, the band width should be greater than 6 eV if band-like holes are to be predicted. The calculated small hole polaron binding energy is $|E_B| \simeq 3$ eV. This value must be compared with the correspondingly large valence band widths which are characteristic of several oxides [325], e.g. MgO (6.5 eV), TiO_2 (5.5 eV), VO_2 (5.6 eV) and $SrTiO_3$ (6.5 eV).

In order to complete our discussion of basic electronic properties in YIG we consider some optical absorption processes related to electron/hole transfer between different ions. For crude estimates we may use shell model defect energies appropriately corrected by ionization and crystal field energies. It is understood that all absorption energies subsequently reported mainly yield qualitative guidance for interpretations and should not be taken quantitatively. This, of course, would demand that we explicitly include the electronic structure by means of exact quantum mechanical procedures as embedded cluster calculations (Sect. 2.1.2). Similar comments also apply to some extent to our investigations related to conductivity, as discussed above. However, in that case, in contrast to the present optical calculations, lattice relaxation effects are fully taken into account and these dominant terms may outweigh the uncertainties due to the inaccuracy in treating the electronic structure. Thus thermal energies are expected to be quantitatively more reliable than optical energies.

Table 7.7 compiles the various optical absorption energies as calculated on the basis of the shell model. All lattice positions of the ion cores have been held fixed during the absorption processes denoted in Table 7.7. The shell model formation energies of iron have been corrected by suitable crystal field splitting energies [320]. All lattice configurations correspond exactly to the initial states. Thus, in all cases initially involving defects, the appropriate lattice geometry is that of the relaxed initial defect configuration. Only ion shells (representing valence electrons) are allowed to move during absorption processes.

The qualitative features of interest may be summarized as follows:

- The charge transfer processes between oxygen and iron ions show similar absorption energies to those between pairs of iron ions.
- The energy gap is determined by the iron pair process:

Table 7.7. Optical absorption processes

Absorption process	Absorption energies (eV)
Electron transfer between iron pairs	
$\mathrm{Fe}^{\times}_{(a)} + \mathrm{Fe}^{\times}_{(d)} \rightarrow \mathrm{Fe}^{\prime}_{(d)} + \mathrm{Fe}^{\bullet}_{(a)}$	11.12
$\mathrm{Fe}^{\times}_{(d)} + \mathrm{Fe}^{\times}_{(a)} \rightarrow \mathrm{Fe}^{\bullet}_{(d)} + \mathrm{Fe}^{\prime}_{(a)}$	7.02
$2\mathrm{Fe}^{\times}_{(a)} \rightarrow \mathrm{Fe}^{\bullet}_{(a)} + \mathrm{Fe}^{\prime}_{(a)}$	11.02
$2\mathrm{Fe}^{\times}_{(d)} \rightarrow \mathrm{Fe}^{\bullet}_{(d)} + \mathrm{Fe}^{\prime}_{(d)}$	9.65
Charge transfer, $O^{2-} \rightarrow Fe^{3+}$	
$\mathrm{Fe}^{\times}_{(a)} + \mathrm{O}^{\times}_{O} \rightarrow \mathrm{Fe}^{\prime}_{(a)} + \mathrm{O}^{\bullet}_{O}$	9.09
$\mathrm{Fe}^{\times}_{(d)} + \mathrm{O}^{\times}_{O} \rightarrow \mathrm{Fe}^{\prime}_{(d)} + \mathrm{O}^{\bullet}_{O}$	9.83
$\mathrm{Fe}^{\times}_{Y} + \mathrm{O}^{\times}_{O} \rightarrow \mathrm{Fe}^{\prime}_{Y} + \mathrm{O}^{\bullet}_{O}$	7.24
Absorption related to surplus electrons and holes	
$\mathrm{Fe}^{\prime}_{(a)} + \mathrm{Fe}^{\times}_{(a)} \rightarrow \mathrm{Fe}^{\times}_{(a)} + \mathrm{Fe}^{\prime}_{(a)}$	2.67
$\mathrm{Fe}^{\prime}_{(a)} + \mathrm{Fe}^{\times}_{(d)} \rightarrow \mathrm{Fe}^{\times}_{(a)} + \mathrm{Fe}^{\prime}_{(d)}$	3.06
$\mathrm{Fe}^{\prime}_{Y} + \mathrm{Fe}^{\times}_{(d)} \rightarrow \mathrm{Fe}^{\times}_{Y} + \mathrm{Fe}^{\prime}_{(d)}$	4.10
$\mathrm{Fe}^{\prime}_{Y} + \mathrm{Fe}^{\times}_{(a)} \rightarrow \mathrm{Fe}^{\times}_{Y} + \mathrm{Fe}^{\prime}_{(a)}$	3.60
$\mathrm{Fe}^{\bullet}_{(d)} + \mathrm{Fe}^{\times}_{(d)} \rightarrow \mathrm{Fe}^{\times}_{(d)} + \mathrm{Fe}^{\bullet}_{(d)}$	5.78
$\mathrm{Fe}^{\bullet}_{(d)} + \mathrm{Fe}^{\times}_{(a)} \rightarrow \mathrm{Fe}^{\times}_{(d)} + \mathrm{Fe}^{\bullet}_{(a)}$	8.32

$$\mathrm{Fe}^{\times}_{(a)} + \mathrm{Fe}^{\times}_{(d)} \longrightarrow \mathrm{Fe}^{1}_{(a)} + \mathrm{Fe}^{\bullet}_{(d)}.$$

- An absorption band close to the fundamental absorption (gap) is predicted to stem from charge transfer between oxygen and $\mathrm{Fe_Y}$ antisite defects.
- Excess electrons and holes essentially lead to optical absorptions within the gap region. Hole transfer needs more energy than electron transfer. The ordering of absorption energies agrees with that corresponding to conductivity activation energies for n- and p-type YIG crystals respectively [3].

Note that the aforementioned assignment of the electronic gap is in line with photoconductivity [326] and electrical conductivity [318] measurements.

7.5 Impurities

The shell model simulations described in the preceding sections have shown that the iron-related antisite defects in particular (i.e. $\mathrm{Fe}^{\times}_{\mathrm{Y}}$ in Kröger–Vink notation [205]) play a major role in YIG crystals. Antisite defects may result from Fe_2O_3 non-stoichiometry, but the actual antisite concentration is significantly larger if interchange reactions (Sect. 7.3) are taken into account. The total number of antisites is expected to be sufficiently large in order to influence other defect formation mechanisms.

In this section we discuss the incorporation of extrinsic (or impurity) defect cations. Divalent, trivalent and tetravalent impurity defects are considered. Whereas trivalent ions substituting for intrinsic cations do not require charge compensating defects for their incorporation, such defects are indispensable for aliovalent ions to maintain the charge neutrality of the YIG lattice. Lead and silicon represent important examples of such aliovalent impurity ions. Special attention will be paid to the particular influence of antisite defects.

Interionic cation–oxygen potential parameters modelling the short-range impurity–lattice interaction have been taken from the work of Lewis and Catlow [122, 158] on oxide crystals, with the exception of the parameters for Pb^{2+}... O^{2-} [327] and Bi^{3+}... O^{2-} [328]. In the case of the rare earth ions the "A" parameters of the impurity–oxygen short-range Buckingham potentials are increased by 150 (eV) in order to reproduce better the observed lattice constants of the corresponding rare earth iron garnets. Table 7.8 lists the calculated lattice constants of various garnets using the original potentials from Lewis et al. and those calculated with the modified parameters. The observed lattice constants are included, too. Table 7.9 compiles all impurity–oxygen potential parameters used in the simulations.

Table 7.8. Simulated lattice constants of some rare earth iron garnets. Details are given in the text

Garnet	Lattice constant (Å)		
	R^{3+}... O^{2-} from Lewis	Corrected R^{3+}... O^{2-}	Exp. value [3]
LuIG	12.09	12.19	12.28
YbIG	12.09	12.23	12.30
GdIG	12.31	12.44	12.47
NdIG	12.45	12.58	12.60
LaIG	12.61	12.73	12.73

Table 7.9. List of all short-range impurity–oxygen potentials used in our simulation study. The potentials were cut off at a distance from the impurity cations of 13.6 Å

Interaction	A (eV)	ϱ (Å)
Mg^{2+}... O^{2-}	1428.5	0.2945
Ca^{2+}... O^{2-}	1090.4	0.3437
Pb^{2+}... O^{2-}	5444.4	0.2994
Lu^{3+}... O^{2-}	1497.1	0.3430
Yb^{3+}... O^{2-}	1459.6	0.3462
Gd^{3+}... O^{2-}	1486.8	0.3551
Ho^{3+}... O^{2-}	1500.2	0.3487
Eu^{3+}... O^{2-}	1508.0	0.3556
Nd^{3+}... O^{2-}	1529.9	0.3601
La^{3+}... O^{2-}	1589.7	0.3651
Bi^{3+}... O^{2-}	11059.7	0.2751
Al^{3+}... O^{2-} 1.	1474.4	0.3006
Al^{3+}... O^{2-} 2.	1114.9	0.3118
Mn^{3+}... O^{2-}	1257.9	0.3214
Sc^{3+}... O^{2-}	1299.4	0.3312
Si^{4+}... O^{2-}	913.2	0.3428
Zr^{4+}... O^{2-}	1608.1	0.3509
Th^{4+}... O^{2-}	1147.7	0.3949

7.5.1 Incorporation of Trivalent Impurities

The general solution mechanism for trivalent impurity cations is given by the reaction

$$\frac{1}{2}M_2O_3 +' YIG' \longrightarrow M_C^{\times} + \frac{1}{2}C_2O_3\,. \qquad (7.27)$$

In (7.27) 'YIG' means the bulk crystal, M the specified trivalent impurity ion and C one of the replaced intrinsic cations in YIG, i.e. Y, $Fe_{(a)}$ or $Fe_{(d)}$. Due to the absence of charge misfit effects the ionic radii are most decisive in determining the lattice site at which the impurity is incorporated. Ion-size effects are described accurately by means of the appropriate short-range impurity–oxygen pair potentials. According to (7.27) the defect formation energy of $M_C^{\times}$ measures the ion-size misfit between the cations M and C. Moreover, in order to calculate the preferred incorporation site of an impurity, e.g. Y and $Fe_{(a)}$, one must compare the corresponding difference in the defect energy with half the difference of the respective C_2O_3 lattice energies, which to some extent reflects the different sizes of the intrinsic cations. Table 7.10 summarizes the theoretically predicted and experimentally found incorporation sites of various trivalent cations. The site stabilization energy in Table 7.10 is defined as the energy difference according to (7.27) for the two most favourable incorporation sites.

Table 7.10. Predicted (•) and observed (◇) incorporation sites for trivalent impurity cations. In addition, the calculated next favoured site is indicated by a "+". The predictions refer to shell model simulations of defects in otherwise pure YIG

Impurity cation	Substituted intrinsic cation Y^{3+}	$Fe^{3+}_{(a)}$	$Fe^{3+}_{(d)}$	Site stabilization energy (eV)	Impurity Ion radius (Å)[a]
Al^{3+}		•	+ ◇	0.17	0.4
Cr^{3+}		• ◇	+	1.50	0.615
Mn^{3+}		• ◇	+	0.90	0.645
Sc^{3+}		• ◇	+	1.14	0.745
Lu^{3+}	• ◇	+		0.38	0.972
Yb^{3+}	• ◇	+		0.43	0.982
Ho^{3+}	• ◇	+		0.59	1.02
Gd^{3+}	• ◇	+		0.77	1.061
Eu^{3+}	• ◇	+		0.81	1.073
Nd^{3+}	• ◇	+		0.97	1.12
Bi^{3+}	• ◇	+		2.10	1.132
La^{3+}	• ◇	+		1.16	1.19

[a] R. D. Shannon and C. T. Prewitt. *Acta Cryst.*, B25:925, 1969

In the case of Cr^{3+} and Mn^{3+} the shell model site stabilization energies have been corrected by (cubic) crystal field energy terms in order to take electronic open-shell effects for these 3d-transition metal ions into account. Crystal field energies are estimated using the expressions

$$\Delta E_{\mathrm{CF}}^{\mathrm{oct}} = \frac{3}{5} n_{\mathrm{e_g}} (10\mathrm{Dq})_{\mathrm{oct}} - \frac{2}{5} n_{\mathrm{t_{2g}}} (10\mathrm{Dq})_{\mathrm{oct}} \tag{7.28}$$

for octahedrally coordinated transition metal ions and similarly

$$\Delta E_{\mathrm{CF}}^{\mathrm{tet}} = \frac{2}{5} n_{\mathrm{t_{2g}}} (10\mathrm{Dq})_{\mathrm{tet}} - \frac{3}{5} n_{\mathrm{e_g}} (10\mathrm{Dq})_{\mathrm{tet}} \tag{7.29}$$

for cations replacing tetrahedrally coordinated iron ions. For simplicity we assume as a crude approximation $(10\mathrm{Dq})_{\mathrm{oct}} \approx (10\mathrm{Dq})_{\mathrm{tet}}$ with $(10\mathrm{Dq})_{\mathrm{oct}} \sim$ 2eV [329]. In using this value we overestimate the tetrahedral crystal field contributions, which, in turn, leads to a slight underestimation of the now pronounced octahedral site stabilization energies for Cr^{3+} and Mn^{3+}. $n_{\mathrm{e_g}}$ and $n_{\mathrm{t_{2g}}}$ denote the electronic occupation numbers of the $\mathrm{e_g}$ and $\mathrm{t_{2g}}$ one-electron states, respectively. The predictions for these two transition metal ions do not depend critically upon the inclusion of such crystal field corrections, since the uncorrected octahedral site stabilization energies (which then represent lower bounds) are 0.7 eV for Cr^{3+} and 0.5 eV for Mn^{3+}. Crystal field corrections have been neglected in the case of open-shell rare earth ions, because the corresponding energy terms are expected to be small due to the shielding of 4f electrons against the embedding crystal field.

Inspecting Table 7.10 we observe the very satisfactory agreement between theory and experiment. Only in the case of Al^{3+} are the predictions uncertain, as there is only a small energy difference between the respective lattice sites. For aluminium an octahedral site preference is predicted; however, the tetrahedral site occupation is only 0.17 eV less favourable, suggesting that the aluminium might be close to a critical ion size where a change of the incorporation in favour of the tetrahedral cation site takes place. In order to investigate this problem further we use a second Al^{3+}... O^{2-} short-range potential suggested by Lewis et al. [122]. This potential is slightly less repulsive than the first one. However, it favours the tetrahedral site occupation by only 0.02 eV. Both Al^{3+}-related short-range potentials must be tested by examining their ability to reproduce the lattice structure and macroscopic properties of yttrium aluminium garnet (YAG). For this purpose the respective aluminium parameters are substituted for the iron parameters in the YIG parameter set. As a result the second potential underestimates the YAG lattice constant by 0.2 Å, whereas the first potential is significantly more accurate in reproducing structure and macroscopic properties of YAG (i.e. it gives a calculated lattice constant of 11.99 Å compared with the observed value of 12.006 Å [3]). Moreover, it is not reasonable to scale "A"–Buckingham parameters appropriate to the various coordination numbers of cations in garnet crystals. The scaling procedure originally proposed by Lewis et al. [122] leads in the case of garnets to less satisfactory simulations of the lattice structure. The present simulations on YIG and YAG suggest that the cation–oxygen short-range potentials for the octahedral and tetrahedral lattice sites are nearly identical. This finding is in agreement with earlier investigations on YAG [316].

In conclusion, the first Al...O potential is better suited for modelling aluminium impurities in YIG. Thus, we would not predict a pronounced tendency for Al^{3+} ions to enter the tetrahedral cation site. In contrast to these calculations, experimental investigations [3] show that Al^{3+} ions occupy the tetrahedral site with occupation fractions $n_{(d)}/(n_{(a)} + n_{(d)}) > 0.8$. When comparing the theoretical predictions with experimental investigations it should be kept in mind, however, that the simulations refer to perfectly grown YIG. On the other hand, most existing YIG crystals are defective to some extent. Many samples are non-stoichiometric (i.e. Fe_2O_3 excess) or contain a considerable amount of different impurity ions, e.g. Bi^{3+}, Pb^{2+} or other big cations which dilate the lattice. The present simulations confirm that the existence of big cations tends to favour the incorporation of Al^{3+} on tetrahedral lattice sites. A simulation of Al^{3+} in $(Y_{\frac{9}{4}}Bi_{\frac{3}{4}})Fe_5O_{12}$ (see also Sect. 7.2) yields a reduced octahedral site stabilization energy of ~0.1 eV using the first Al^{3+}... O^{2-} potential but a pronounced tetrahedral site preference corresponding to a stabilization energy of 0.5 eV in the case of the second short-range potential. Although quantitatively not fully satisfactory, these calculations readily demonstrate the qualitative trend resulting from lattice dilation due to large cations. Similarly, small rare earth cations such as Lu^{3+} would preferentially enter the octahedrally coordinated iron site in garnet crystals with expanded lattices compared with perfect YIG (indeed, calculations show that the dodecahedral site stabilization energy of Lu^{3+} in $(Y_{\frac{9}{4}}Bi_{\frac{3}{4}})Fe_5O_{12}$ is reduced by ~0.1 eV compared with perfect YIG). It is observed in this context that Suchow et al. [330] succeeded in preparing gallium garnets in which the octahedral lattice sites were occupied by Lu^{3+} ions after expanding the whole lattice with large rare earth ions (e.g. Nd^{3+}) on dodecahedral sites. In contrast, the presence of significant concentrations of small cations leading to a lattice contraction of YIG would stabilize the Al^{3+} octahedral site occupation, whereas in the case of small rare earth ions the dodecahedral site preference became more pronounced.

Further influences on impurity solutions may be certain magnetic effects which are neglected within shell model simulations. Crude estimates suggest that such terms are small, but in the case of Al^{3+} these terms may help to increase the tetrahedral site occupancy. Employing the "LCM"-model of Nowik and Rosencwaig [331, 332] according to which a superexchange interaction is assumed to be present only between nearest iron ions, we can estimate a contribution of 0.1 eV from magnetic energy terms favouring the tetrahedral site, which is of the order of the octahedral site preference determined for Al^{3+} in otherwise perfect YIG. This value is calculated from the observation that $Fe_{(d)}$ has four $Fe_{(a)}$ nearest neighbours but that $Fe_{(a)}$, in turn, has six $Fe_{(d)}$ nearest neighbours. Incorporation of diamagnetic cations such as Al^{3+}

diminishes the superexchange energy, which in perfect crystals maintains the observed ground state ferrimagnetic order. The magnetic energy difference between tetrahedral and octahedral site occupation is then given by:

$$\Delta E_{\mathrm{m}} = 4 J_{\mathrm{ad}} S(S+1)\,, \tag{7.30}$$

employing a Heisenberg type Hamiltonian $\mathcal{H}_i = -2J_{\mathrm{ad}} \sum_j \mathbf{S}_i . \mathbf{S}_j$ describing the antiparallel spin coupling between nearest iron neighbours with $J_{\mathrm{ad}} = -0.003$ eV [333] and $S = \frac{5}{2}$.

We conclude that, in order to predict from shell model simulations the correct incorporation of trivalent impurity cations which do not show a definite site preference in perfectly grown YIG, it seems to be necessary to include the precise chemical composition of YIG crystals under investigation as well as small magnetic effects.

7.5.2 Incorporation of Divalent and Tetravalent Impurities

We start our considerations by separately discussing the incorporation of divalent and tetravalent impurity ions. The results are relevant for YIG crystals which dominantly or exclusively contain the respective impurity species. Finally, we shall consider the combined incorporation of both divalent and tetravalent impurity cations in order to single out possible impurity-restricted compensation mechanisms. Such extrinsic charge compensations have often been proposed in the literature (see [3], which also gives additional references).

Divalent impurity ions represent point defects which are negatively charged with respect to the embedding host lattice. Charge compensators must therefore be positively charged. Due to the present calculations the most favourable incorporation reactions for divalent impurity ions are given by

$$2\mathrm{MO} +' \mathrm{YIG}' \longrightarrow 2\mathrm{M}_{\mathrm{Y}}^{\mathrm{l}} + \mathrm{V}_{\mathrm{O}}^{\bullet\bullet} + \mathrm{Y_2O_3} \tag{7.31}$$

$$2\mathrm{MO} +' \mathrm{YIG}' \longrightarrow 2\mathrm{M}_{\mathrm{Y}}^{\mathrm{l}} + 2\mathrm{Y}_{\mathrm{(a)}}^{\times} + \mathrm{V}_{\mathrm{O}}^{\bullet\bullet} + \mathrm{Fe_2O_3} \tag{7.32}$$

$$2\mathrm{MO} +' \mathrm{YIG}' \longrightarrow 2\mathrm{M}_{\mathrm{(a)}}^{\mathrm{l}} + \mathrm{V}_{\mathrm{O}}^{\bullet\bullet} + \mathrm{Fe_2O_3} \tag{7.33}$$

$$2\mathrm{MO} +' \mathrm{YIG}' \longrightarrow 2\mathrm{M}_{\mathrm{(d)}}^{\mathrm{l}} + \mathrm{V}_{\mathrm{O}}^{\bullet\bullet} + \mathrm{Fe_2O_3}\,, \tag{7.34}$$

i.e. irrespective of their incorporation site these impurities are compensated by the formation of oxygen vacancies. Equation (7.32) in addition describes the formation of $\mathrm{Y_{(a)}}$ antisite defects. Table 7.11 summarizes the corresponding site stabilization energies for Mg^{2+}, Ca^{2+} and Pb^{2+}.

The predicted substitution of Mg^{2+} for octahedrally coordinated iron as well as the yttrium site incorporation of Ca^{2+} and Pb^{2+} impurities are in agreement with the corresponding experimental results [3]. Reaction (7.32) is by only 0.6 eV less favourable than (7.31). This reaction energy difference is further reduced if the YIG crystals under investigation contain an

Table 7.11. Oxygen vacancy mediated incorporation of divalent impurity cations. Positive energies mean that the first site is favoured over the second one

Impurity cation	Predicted incorporation site	Site stabilization energy (eV) per MO	
		Y over $Fe_{(a)}$	$Fe_{(a)}$ over $Fe_{(d)}$
Mg^{2+}	$Fe_{(a)}$	−0.50	1.20
Ca^{2+}	Y	0.39	3.02
Pb^{2+}	Y	1.43	0.27

appreciable proportion of large cations such as Pb^{2+} on a Y site or Y^{3+} on octahedral iron sites (resulting from impurity incorporation, for instance). For example, the energy difference in $Y_3(Fe_4Y)O_{12}$ is only half as large as in perfect YIG. These calculations only simulate the effect of $Y_{(a)}$ antisite defects. It is speculated that the additional existence of Pb^{2+} ions could lead to a preference for (7.32) over (7.31). Thus, (7.32) is expected to represent the dominant "high-concentration" incorporation mechanism for divalent impurities on dodecahedral cation sites. The incorporation of big cations leads to a dilation of the crystal lattice and, thus, to an increasing formation of $Y_{(a)}$ antisite defects. As a result, the variation of the lattice constant should be more pronounced than that caused by the impurity cations alone. Indeed, in the case of Pb^{2+} the change in the lattice constant has been reported to be twice as large as expected by incorporation of these impurities ([3] and references therein). Corresponding changes in magnetic properties are in line with the formation of yttrium antisite defects. Similar effects have also been observed in other garnet crystals [334, 335].

Tetravalent impurity cations which are positively charged with respect to the lattice are compensated by negatively charged defects, such as cation vacancies, excess electrons or some intrinsic defect clusters with negative net charge. Since in the present context we neglect the existence of other aliovalent impurity species, we are obliged to consider intrinsic charge compensators. The most energetically favourable incorporation reactions involving vacancies are given by:

$$MO_2 +' \mathrm{YIG}' \longrightarrow M^{\bullet}_{(a)} + V^{\bullet\bullet}_{O} + V^{3\prime}_{Fe(d)} + Fe_2O_3 \tag{7.35}$$

$$MO_2 +' \mathrm{YIG}' \longrightarrow M^{\bullet}_{(d)} + V^{\bullet\bullet}_{O} + V^{3\prime}_{Fe(d)} + Fe_2O_3 \tag{7.36}$$

$$3MO_2 +' \mathrm{YIG}' \longrightarrow 3M^{\bullet}_{(a)} + V^{3\prime}_{Fe(d)} + 2Fe_2O_3 \tag{7.37}$$

$$3MO_2 +' \mathrm{YIG}' \longrightarrow 3M^{\bullet}_{(a)} + V^{3\prime}_{Y} + YFeO_3 + Fe_2O_3 \tag{7.38}$$

$$3MO_2 +' \mathrm{YIG}' \longrightarrow 3M^{\bullet}_{(d)} + V^{3\prime}_{Fe(d)} + 2Fe_2O_3 \tag{7.39}$$

$$3MO_2 +' \mathrm{YIG}' \longrightarrow 3M^{\bullet}_{(d)} + V^{3\prime}_{Y} + YFeO_3 + Fe_2O_3 \tag{7.40}$$

$$3MO_2 +' \mathrm{YIG}' \longrightarrow 3M^{\bullet}_{Y} + V^{3\prime}_{Y} + 2Y_2O_3 \tag{7.41}$$

$$3MO_2 +' YIG' \longrightarrow 3M_Y^{\bullet} + V_{Fe(d)}^{3l} + YFeO_3 + Y_2O_3 \,. \tag{7.42}$$

Equations (7.35) and (7.36) show an example of suitable compensating defect clusters, i.e. $(V_O + V_{Fe(d)})^{ll}$ (it is noted that both vacancy partners are not located on nearest lattice sites, since this configuration would be unstable, i.e. 1.1 eV less favourable than separated vacancies). In the case of impurity incorporation on iron sites (7.37) and (7.39) are by 5.2 eV (per molecule MO_2) more favourable than (7.35) and (7.36), respectively. Consequently, we may conclude that $(V_O + V_{Fe(d)})^{ll}$ clusters do not play a significant role as charge compensators of tetravalent impurity ions. Similarly, one can rule out $(V_O+V_Y)^{ll}$ defect clusters. Thus, single cation vacancies are significantly preferred as charge compensators. Since reaction (7.38) is by only 0.24 eV (per MO_2) less favourable than (7.37) charge compensating iron vacancies are only slightly preferred against alternative yttrium vacancies. Y site incorporation, on the other hand, is accommodated by reaction (7.42) which is by 0.47 eV per MO_2 more favourable than (7.41). We conclude that iron vacancies seem to represent the majority of charge compensating cation vacancies accompanying the incorporation of tetravalent impurity cations. As the formation of $YFeO_3$ is energetically favourable[1], yttrium vacancies may to some extent take part in compensating tetravalent cations on iron sites. We emphasize that the majority of assumed iron vacancies refer to the tetrahedral iron site, since the formation of $V_{Fe(a)}^{3l}$ vacancies is by 0.6 eV less favourable (see Sect. 7.3). Table 7.12 compiles the site stabilization energies for Si^{4+}, Zr^{4+} and Th^{4+} cations.

The predictions for these impurity ions are in agreement with corresponding experimental investigations [3]. However, in the case of Zr^{4+} the calculations suggest an almost equal distribution of these cations between dodecahedral and octahedral sites, as the site stabilization energy favouring the octahedral site is not very significant. Subsequently we will see that these predicted site occupations remain qualitatively unaltered if charge compensation is given by divalent impurities instead of intrinsic defects. Moreover, the octahedral site stabilization energy for Zr^{4+} increases if compensating divalent extrinsic ions occupying the yttrium site are involved. Before dis-

---

[1] In order to obtain reliable defect chemical predictions one must use the energetically most favourable structure simulating a crystal corresponding to the chemical formula $YFeO_3$. According to the present potential model for YIG (set 2 in Sect. 7.1) this structure corresponds to Y-rich garnet. However, the discussion in the preceding sections also indicated that $YFeO_3$ with the perovskite structure could instead be slightly more favourable. This would fit with the observation that there is no YIG with excess yttrium. The shell model simulations are ultimately inconclusive in this respect, since small potential variations turned out to be sufficient to reverse the results. This uncertainty does not influence the present defect calculations seriously. The only effects would consist of a slightly more pronounced role of yttrium vacancies and of an increased octahedral site stabilization energy for Zr^{4+} when compensated by divalent cations on yttrium sites.

Table 7.12. Cation vacancy mediated incorporation of tetravalent impurity cations. Positive energies mean the first site is favoured over the second one

Impurity cation	Predicted incorporation site	site stabilization energy (eV) per MO_2 Y over $Fe_{(a)}$	$Fe_{(a)}$ over $Fe_{(d)}$
Si^{4+}	$Fe_{(d)}$	−6.16	−3.2
Zr^{4+}	$Fe_{(a)}$	−0.06	4.2
Th^{4+}	Y	0.47	5.7

cussing the combined incorporation of divalent and tetravalent cations we consider charge-compensating electrons as an alternative to the formation of compensating cation vacancies. The corresponding reactions are:

$$MO_2 + ' YIG' \longrightarrow M^{\bullet}_{Fe} + Fe^{l}_{(a)} + \frac{1}{2}Fe_2O_3 + \frac{1}{4}O_2(g) , \quad (7.43)$$

for incorporation on iron sites and

$$MO_2 + ' YIG' \longrightarrow M^{\bullet}_{Y} + Fe^{l}_{(a)} + \frac{1}{2}Y_2O_3 + \frac{1}{4}O_2(g) \quad (7.44)$$

describing the substitution for Y^{3+} cations. Analogous equations may be formulated by assuming the presence of $Fe^{\times}_{Y}$ antisite defects, e.g.:

$$MO_2 + Fe^{\times}_{Y} + ' YIG' \longrightarrow M^{\bullet}_{Fe} + Fe^{l}_{Y} + \frac{1}{2}Fe_2O_3 + \frac{1}{4}O_2(g) . \quad (7.45)$$

The calculations show that electronic compensation is as favourable as vacancy-assisted impurity incorporation only if Fe_Y antisite defects are assumed to exist. The reaction energy difference between (7.45) and (7.43), given by $\Delta E = (E(Fe^{l}_{Y}) - E(Fe^{\times}_{Y})) - E(Fe^{l}_{(a)})$, amounts to −1.15 eV per molecule MO_2. The incorporation mechanisms with vacancy and electronic Fe^{l}_{Y} charge compensation differ in energy by only 0.4 eV. Additional crystal field splitting terms for Fe^{l}_{Y} neglected so far would favour reaction (7.45) over (7.39), if we assume a typical crystal field splitting energy of 10Dq $\sim$ 1–2 eV. There are experimental indications in favour of the electronic compensation: in Si-doped YIG single crystals a one-to-one correspondence between the Si and ferrous concentration has been observed [310]. However, there is a maximum Si concentration (x^{Si}_{max} ∼0.1 Si per formula unit [336]) up to which this correspondence holds. For larger Si concentrations a vacancy-type charge compensation has been suggested. These experimental findings fit with reaction mechanisms (7.39) and (7.45), with (7.45) being slightly more favourable than (7.39). According to this model the maximum Si content with electronic compensation is determined by the concentration of Fe_Y antisite defects. The suggested value for x^{Si}_{max} is in agreement with this interpretation. x^{Si}_{max} may be slightly larger than the maximum concentration of antisite defects since, in addition, some Si ions should be charge-compensated by certain divalent impurities which are always present in YIG. We now consider in further detail the possibility of such charge compensation mechanisms.

If, for example, divalent and tetravalent impurity species are present in YIG crystals with comparable concentrations one could imagine the possibility of charge compensation among the impurity cations (extrinsic compensation). A typical incorporation mechanism is given by

$$\mathrm{MO} + \mathrm{NO_2} +' \mathrm{YIG}' \longrightarrow \mathrm{M}^{\prime}_{\mathrm{Y}} + \mathrm{N}^{\bullet}_{\mathrm{(d)}} + \mathrm{YFeO_3}\,, \tag{7.46}$$

where M and N denote divalent and tetravalent impurity ions, respectively. It is straightforward to formulate further incorporation reactions with different choices for the substitution sites. By taking the appropriate differences of the separate impurity incorporation reactions given by (7.31)–(7.42) and the extrinsic compensation mechanisms (such as (7.46), for example) we can determine which is the most energetically favoured of the latter type of reactions. The relevant energy differences are independent of the respective impurity defect energies. The detailed energy expressions $\Delta E(\mathrm{M}^{\prime}_{\alpha}, \mathrm{N}^{\bullet}_{\beta})$ describing a divalent cation M on site α and a tetravalent cation N on site β are given as follows:

$$\Delta E(\mathrm{M}^{\prime}_{\alpha}, \mathrm{N}^{\bullet}_{\beta}) = \frac{1}{2} E(\mathrm{V}^{\bullet\bullet}_{\mathrm{O}}) + \frac{1}{3} E(\mathrm{V}^{3\prime}_{\mathrm{Fe(d)}}) + \sum_i f_i^{\alpha,\beta} E^{i}{}_{\mathrm{Lat}}\,, \tag{7.47}$$

where $f_i^{\alpha,\beta}$ are rational numbers, depending on sites α and β, to be multiplied with the lattice energy E^i_{Lat} of the oxide i (=Y_2O_3, Fe_2O_3, $YFeO_3$). The calculated energies are summarized in Table 7.13.

For all possible combinations of impurity site occupations the extrinsic compensation is more favourable than incorporation mechanisms involving intrinsic charge compensators (the alternative assumption of compensating electrons, i.e. $\mathrm{Fe}^{\prime}_{\mathrm{Y}}$, instead of vacancies would not change the results qualitatively). This result is in excellent agreement with the large number of observed impurity compensations [3]. As a further consequence of the above considerations we conclude that the incorporation sites of impurity ions remain

Table 7.13. Comparison of intrinsic and extrinsic charge compensations for divalent and tetravalent cations. Positive energies mean that the extrinsic compensation solution mode is favoured over the alternative separate incorporation by means of intrinsic vacancies

Impurity species $\mathrm{M}^{\prime}_{\alpha}, \mathrm{N}^{\bullet}_{\beta}$	$\Delta E(\mathrm{M}^{\prime}_{\alpha}, \mathrm{N}^{\bullet}_{\beta})$ (eV)
$\mathrm{M}^{\prime}_{\mathrm{Y}}, \mathrm{N}^{\bullet}_{\mathrm{Y}}$	2.46
$\mathrm{M}^{\prime}_{\mathrm{Y}}, \mathrm{N}^{\bullet}_{\mathrm{Fe}}$	2.59
$\mathrm{M}^{\prime}_{\mathrm{Fe}}, \mathrm{N}^{\bullet}_{\mathrm{Fe}}$	2.59
$\mathrm{M}^{\prime}_{\mathrm{Fe}}, \mathrm{N}^{\bullet}_{\mathrm{Y}}$	2.59

unchanged compared with those found with intrinsic charge compensators. This can be seen by comparing the site stabilization energies listed in Table 7.12 with corresponding energy differences in Table 7.13. In particular, it is observed that charge-compensating divalent cations on the yttrium site lead to an increased octahedral site stabilization energy of 0.2 eV for Zr^{4+}.

Finally, we briefly consider the formation of Pb^{4+} ions. Since there is not currently an appropriate Pb^{4+}... O^{2-} short-range potential, we can approximate this by taking the Zr^{4+}... O^{2-} potential. This procedure is justified as a crude approximation by the observation that both ion species are of nearly the same size and, in addition, do not possess an open electronic (sub)shell. Since the usual melt compositions from which YIG crystals are grown contain PbO, we should assume that lead ions naturally occur as divalent cations in YIG and that the formation of Pb^{4+} involves oxidation taking place during the crystal growth. Inspection of Table 7.12 shows that Pb^{4+} ions prefer to occupy the octahedral iron site. Possible chemical reactions governing the formation of Pb^{4+} ions in YIG are given by

$$2\mathrm{PbO} + \frac{1}{2}\mathrm{O}_2(\mathrm{g}) +' \mathrm{YIG}' \longrightarrow \mathrm{Pb}^{1}_{\mathrm{Y}} + \mathrm{Pb}^{\bullet}_{(\mathrm{a})} + \mathrm{FeYO}_3 \tag{7.48}$$

$$2\mathrm{PbO} + \frac{1}{2}\mathrm{O}_2(\mathrm{g}) +' \mathrm{YIG}' \longrightarrow \mathrm{Pb}^{1}_{\mathrm{Y}} + \mathrm{Pb}^{\bullet}_{\mathrm{Y}} + \mathrm{Y}_2\mathrm{O}_3\,. \tag{7.49}$$

Note that an evaluation of the reaction energies of (7.48) and (7.49) involves the third and fourth ionization potentials of Pb atoms (I^3_{Pb}=31.9 eV, I^4_{Pb}=42.1 eV [337]). The calculated reaction energies per molecule PbO are 0.46 eV and 0.74 eV for (7.48) and (7.49), respectively. These results indicate that tetravalent lead ions will occupy both the dodecahedral and the octahedral cation sites with slight preference for the octahedral site. Moreover, the predicted reaction energies suggest that Pb^{4+} ions could be present in YIG at significant concentrations. However, whereas the Pb^{2+}/Pb^{4+} mechanisms are endothermic (note also that the incorporation of gaseous oxygen corresponds to a negative entropy change leading to $\Delta G > \Delta E$, as was already mentioned in Sect. 7.4), the alternative lead incorporation reactions discussed above are exothermic. It thus follows that Pb^{4+} ions represent a minority charge species compared with divalent lead. Of course, the actual Pb^{4+} concentration depends on the external oxygen pressure applied to the system. From (7.48) and (7.49) it is obvious, however, that even the highest oxygen pressures would result in equal concentrations of divalent and tetravalent lead ions. "Complete" oxidation type lead incorporation reactions in which lead ions exclusively occur as tetravalent cations are given by

$$3\mathrm{PbO} + \frac{3}{2}\mathrm{O}_2(\mathrm{g}) +' \mathrm{YIG}' \longrightarrow 3\mathrm{Pb}^{\bullet}_{(\mathrm{a})} + \mathrm{V}^{31}_{\mathrm{Fe(d)}} + 2\mathrm{Fe}_2\mathrm{O}_3 \tag{7.50}$$

$$3\mathrm{PbO} + \frac{3}{2}\mathrm{O}_2(\mathrm{g}) +' \mathrm{YIG}' \longrightarrow 3\mathrm{Pb}^{\bullet}_{\mathrm{Y}} + \mathrm{V}^{31}_{\mathrm{Fe(d)}} + \mathrm{YFeO}_3 + \mathrm{Y}_2\mathrm{O}_3 \tag{7.51}$$

with calculated reaction energies of 4.74 eV and 4.80 eV per PbO, respectively. These results suggest that it is impossible to achieve complete oxidation of lead ions in YIG.

8. Summary and Conclusions

This volume reviews atomistic simulations of complex oxides as used in many electro- and magnetooptic applications. Generally, corresponding materials are characterized by bonding properties intermediate between covalent and ionic bonding; covalency effects are significantly intermixed with lattice relaxations, giving rise to what we may call semi-ionicity. This general behaviour becomes particularly obvious in defect formation. It should be accounted for in any theoretical simulation study. We recall, that defects are responsible for many technological material properties.

Atomistic crystal properties are usually investigated by – sometimes expensive – experiments and by suitable theoretical simulations. In many instances such "computer experiments" provide a reasonable and reliable guide to interpretations of experimental data. Whereas the atomistic properties of metals and semiconducting materials can be inferred solely from state-of-the-art electronic structure calculations, ionic crystals, on the other hand, may be modelled reliably by means of potential simulations, of which the shell model parametrized in terms of suitable pair potentials is the cheapest and most flexible one that accounts for important long-range lattice deformations. Further, the electronic properties of ionic materials closely resemble well-understood free-ion properties, and may be described by an embedded atom approach. Simulations of semi-ionic systems require various approaches, ranging from all-electron supercell calculations to pure potential simulations in order to account for the intermediate nature of these materials. Of particular importance are embedded cluster calculations, which may be considered a hybrid between electronic structure calculations and potential simulations. They have been found to be well suited to studying defects and their local electronic structure properties taking long-range defect-induced lattice deformations and covalency effects into account self-consistently. Many defect simulations of this type have been performed during the recent years for perovskite-structured oxides ABO_3, but more complex systems like SBN and YIG are still awaiting their embedded cluster simulations. The situation is similar for $LiNbO_3$. These investigations should be part of forthcoming research programmes. At the same time, the high quality of present simulation techniques could be increased slightly further by enlarging the quantum clusters in the case of embedded cluster calculations and by including lattice relaxations beyond nearest-neighbour atom shells of defects within supercell simulations. However, the latter task requires the use of very large supercells in order to avoid any spurious interactions between different supercells.

References

1. Y. Fainman, J. Ma and S. H. Lee. *Materials Science Reports*, 9:53, 1993.
2. P. Günter and J.-P. Huignard, editors. *Photorefractive Materials and their Applications I and II*, volumes 61 & 62 of *Topics in Applied Physics*. Springer Berlin Heidelberg, 1988.
3. G. Winkler. *Magnetic Garnets*. Vieweg Tracts in Pure and Applied Physics. Vieweg Verlag, Braunschweig, Wiesbaden, 1981.
4. H. Hemme, H. Dötsch and H. P. Menzler. *Appl. Optics*, 26:3811, 1987.
5. H. Hemme, H. Dötsch and P. Hertel. *Appl. Optics*, 29:2741, 1990.
6. P. Labbe. *Key Engineering Materials*, 68:293, 1992.
7. F. D. M. Haldane and P. W. Anderson. *Phys. Rev.*, B13:2553, 1976.
8. A. S. Alexandrov and N. F. Mott. *Rep. Prog. Phys.*, 57:1197, 1994.
9. C. Schlenker. In D. Adler, H. Fritzsche and S. R. Ovshinsky, editors, *Physics of Disordered Materials: Mott Festschrift*, volume 3. Plenum Press, New York, 1985.
10. J. H. Harding. *Rep. Prog. Phys.*, 53:1403, 1990.
11. A. R. West. *Solid State Chemistry and its Applications*. John Wiley & Sons, Chichester, 1984.
12. C. R. A. Catlow, C. M. Freeman, M. S. Islam, R. A. Jackson, M. Leslie and S. M. Tomlinson. *Phil. Mag.*, A58:123, 1988.
13. R. Car and M. Parrinello. *Phys. Rev. Lett.*, 55:2471, 1985.
14. M. J. Gillan. *J. Chem. Soc., Faraday Trans. 2*, 85:521, 1989.
15. M. J. Gillan. *J. Phys.: Condens. Matter*, 1:689, 1989.
16. P. Fulde. *Electron Correlations in Molecules and Solids*, volume 100 of Solid-State Sciences. Springer, Berlin Heidelberg, 1991.
17. A. Szabo and N. S. Ostlund. *Modern Quantum Chemistry – Introduction to Advanced Electronic Structure Theory*. McGraw-Hill, New York, 1989.
18. J. C. Slater. *The Self-consistent Field for Molecules and Solids*, volume 4. McGraw-Hill, New York, 1974.
19. A. M. Stoneham. *Theory of Defects in Solids*. Clarendon Press, Oxford, 1975.
20. W. A. Harrison. *Electronic Structure and the Properties of Solids*. W. H. Freeman and Company, San Francisco, 1980.
21. W. R. Hay and P. J. Wadt. *J. Chem. Phys.*, 82:270–310, 1985.
22. W. J. Stevens, H. Basch and M. Krauss. *J. Chem. Phys.*, 81:6026, 1984.
23. R. McWeeny. *Methods of Molecular Quantum Mechanics*. Academic Press, London, 1989.
24. R. O. Jones and O. Gunnarsson. *Rev. Mod. Phys.*, 61:689, 1989.
25. A. L. Fetter and J. D. Walecka. *Quantum Theory of Many-Particle Systems*. McGraw-Hill, New York, 1971.
26. L. Hedin. *Phys. Rev.*, A139:796, 1965.
27. P. Hohenberg and W. Kohn. *Phys. Rev.*, B136:864, 1964.
28. M. Levy. *Proc. Natl. Acad. Sci. (USA)*, 76:6062, 1979.

29. W. Kohn and L. J. Sham. *Phys. Rev.*, A140:1133, 1965.
30. R. G. Parr and W. Yang. *Density-Functional Theory of Atoms and Molecules.* Oxford University Press, Oxford, 1989.
31. L. Hedin and B. I. Lundqvist. *J. Phys. C: Solid State Phys.*, 4:2064, 1971.
32. U. von Barth and L. Hedin. *J. Phys. C: Solid State Phys.*, 5:1629, 1972.
33. D. M. Ceperly and B. J. Alder. *Phys. Rev. Lett.*, 45:566, 1980.
34. A. D. Becke. *Phys. Rev.*, A38:3098, 1988.
35. C. Lee, W. Yang and R. G. Parr. *Phys. Rev.*, B37:785, 1988.
36. R. Colle and D. Salvetti. *Theor. Chim. Acta*, 37:329, 1975.
37. C. Filippi, X. Gonze and C. J. Umrigar. Generalized gradient approximations to density functional theory: Comparison with exact results. In J. M. Seminario, editor, *Recent Developments and Applications of Density Functional Theory.* Elsevier, Amsterdam, 1996.
38. R. M. Dreizler and E. K. U. Gross. *Density Functional Theory – An Approach to the Quantum Many-Body Problem.* Springer, Berlin Heidelberg, New York, 1990.
39. J. P. Perdew and M. Levy. *Phys. Rev.*, B31:6264, 1985.
40. X. Gonze, Ph. Ghosez and R. W. Godby. *Phys. Rev. Lett.*, 74:4035, 1995.
41. E. A. Hylleraas. *Z. Phys.*, 65:209, 1930.
42. Ph. Ghosez, X. Gonze and R. W. Godby. *cond-mat/9706279*, 1997.
43. Z. H. Levine and D. C. Allan. *Phys. Rev. Lett.*, 63:1719, 1989.
44. O. K. Andersen. *Phys. Rev.*, B12:3060, 1975.
45. S.-H. Wei and H. Krakauer. *Phys. Rev. Lett.*, 55:1200, 1985.
46. M. Methfessel. *Phys. Rev.*, B38:1537, 1988.
47. T. Neumann. PhD thesis, University of Osnabrück, 1994.
48. A. V. Postnikov, T. Neumann, G. Borstel and M. Methfessel. *Phys. Rev.*, B48:5910, 1993.
49. R. D. King-Smith and D. Vanderbilt. *Ferroelectrics*, 136:85, 1992.
50. K. Schönhammer, O. Gunnarsson and R. M. Noack. *Phys. Rev.*, B52:2504, 1995.
51. U. Schönberger and F. Aryasetiawan. *Phys. Rev.*, B52:8788, 1995.
52. W. Y. Ching, Zong-Quan Gu and Yong-Nian Xu. *Phys. Rev.*, B50:1992, 1994.
53. Jian Chen, Z. H. Levine and J. W. Wilkins. *Phys. Rev.*, B50:11514, 1994.
54. G. Makov and M. C. Payne. *Phys. Rev.*, B51:4014, 1995.
55. A. V. Postnikov, T. Neumann and G. Borstel. *Ferroelectrics*, 164:101, 1995.
56. A. J. Fisher. *J. Phys. C: Solid State Phys.*, 21:3229, 1988.
57. M. O. Selme and P. Pecheur. *J. Phys. C: Solid State Phys.*, 16:2559, 1983.
58. M. O. Selme, P. Pecheur and G. Toussaint. *J. Phys. C: Solid State Phys.*, 17:5185, 1984.
59. M. O. Selme and P. Pecheur. *J. Phys. C: Solid State Phys.*, 18:551, 1985.
60. A. V. Fisenko, S. A. Prosandeyev and V. P. Sachenko. *phys. stat. sol. (b)*, 137:187, 1986.
61. S. A. Prosandeyev, N. M. Teslenko and A. V. Fisenko. *J. Phys.: Condens. Matter*, 5:9327, 1993.
62. S. A. Prosandeyev and I. A. Osipenko. *phys. stat. sol. (b)*, 192:37, 1995.
63. L. F. Mattheiss. *Phys. Rev.*, B6:4718, 1972.
64. P. O. Löwdin. *J. Chem. Phys.*, 19:1396, 1951.
65. G. A. Baraff and M. Schlüter. *J. Phys. C: Solid State Phys.*, 19:4383, 1986.
66. C. Pisani. *Theor. Chim. Acta*, 72:277, 1987.
67. C. Pisani, R. Dovesi, R. Nada and L. N. Kantorovich. *J. Chem. Phys.*, 92:7448, 1990.
68. C. Pisani, R. Orlando and R. Nada. *Reviews of Solid State Science*, 5:177, 1991.

69. L. É. Bar'yudin. *Sov. Phys. Solid State*, 33:1820, 1991.
70. R. A. Evarestov and V. P. Smirnov. *Site Symmetry in Crystals – Theory and Applications*, volume 108 of Solid-State Sciences. Springer, Berlin Heidelberg, 1993.
71. A. B. Kunz and D. L. Klein. *Phys. Rev.*, B17:4614, 1978.
72. J. M. Vail. *J. Phys. Chem. Solids*, 51:589, 1990.
73. S. Huzinaga and A. A. Cantu. *J. Chem. Phys.*, 55:5543, 1971.
74. L. N. Kantorovich. *J. Phys. C: Solid State Phys.*, 21:5041–5073, 1988.
75. Z. Barandiarán and L. Seijo. *J. Chem. Phys.*, 89:5739, 1988.
76. N. F. Mott and M. J. Littleton. *Trans. Faraday Soc.*, 34:485, 1938.
77. F. M. Michel-Calendini, H. Chermette and J. Weber. *J. Phys. C: Solid State Phys.*, 13:1427, 1980.
78. A. Zunger. Electronic structure of 3d transition-atom impurities in semiconductors. In H. Ehrenreich and D. Turnbull, editors, *Solid State Physics*, volume 39. Academic Press, Orlando (Florida), 1986.
79. Z. Barandiarán and L. Seijo. Embedded-cluster calculations. In *Defects and Disorder in Crystalline and Amorphous Solids*, volume 418 of NATO ASI Series C, page 341. Kluwer Academic Publishers, Dordrecht, Boston, London, 1994.
80. R. W. Grimes, C. R. A. Catlow and A. L. Shluger, editors. *Quantum Mechanical Cluster Calculations in Solid State Studies.* World Scientific, Singapore, 1992.
81. W. H. Press, B. P. Flannery, S. A. Teukolsky and W. T. Vetterling. *Numerical Recipes – The Art of Scientific Computing.* Cambridge University Press, Cambridge, 1986.
82. J. M. Vail, A. H. Harker, J. H. Harding and P. Saul. *J. Phys. C: Solid State Phys.*, 17:3401, 1984.
83. Jun Zuo, R. Pandey and A. B. Kunz. *Phys. Rev.*, B44:7187, 1991.
84. A. M. Woods, R. S. Sinkovits, J. C. Charpie, W. L. Huang, R. H. Bartram and A. R. Rossi. *J. Phys. Chem. Solids*, 54:543, 1993.
85. A. M. Woods, R. S. Sinkovits and R. H. Bartram. *J. Phys. Chem. Solids*, 55:91, 1994.
86. T. J. Gryk and R. H. Bartram. *J. Phys. Chem. Solids*, 56:863, 1995.
87. M. F. Guest and V. R. Saunders. *Mol. Phys.*, 29:873, 1975.
88. T. H. Hillier. *Pure Appl. Chem.*, 51:2183, 1979.
89. A. Veillard and J. Demuynk. In H. F. Schaeffer, editor, *Modern Theoretical Chemistry*, volume 4. Plenum Press, New York, 1977.
90. J. F. Janak. *Phys. Rev.*, B18:7165, 1978.
91. P. Moretti and F. M. Michel-Calendini. *Phys. Rev.*, B34:8538, 1986.
92. K. M. Rabe and U. V. Waghmare. *Phys. Rev.*, B52:13236, 1995.
93. A. M. Stoneham, V. T. B. Torres, P. M. Masri and H. R. Schober. *Phil. Mag.*, A58:93, 1988.
94. M. Scheffler and J. Dabrowski. *Phil. Mag.*, A58:107, 1988.
95. R. Taylor. *Physica*, B131:103, 1985.
96. A. E. Carlsson. Beyond pair potentials in elemental transition metals and semiconductors. In H. Ehrenreich and D. Turnbull, editors, *Solid State Physics*, volume 43. Academic Press, New York, 1990.
97. M. S. Islam, M. Leslie, S. M. Tomlinson and C. R. A. Catlow. *J. Phys. C: Solid State Phys.*, 21:L109, 1988.
98. C. R. A. Catlow, S. M. Tomlinson, M. S. Islam and M. Leslie. *J. Phys. C: Solid State Phys.*, 21:L1085, 1988.
99. N. L. Allan and W. C. Mackrodt. *Phil. Mag.*, A58:555, 1988.
100. M. S. Islam and R. C. Baetzold. *Phys. Rev.*, B40:10926, 1989.

101. R. C. Baetzold. *Phys. Rev.*, B42:56, 1990.
102. M. S. Islam. *Supercond. Sci. Technol.*, 3:531, 1990.
103. M. S. Islam and C. Ananthamohan. *Phys. Rev.*, B44:9492, 1991.
104. M. S. Islam and R. C. Baetzold. *J. Mater. Chem.*, 4:299, 1994.
105. M. S. Islam and L. J. Winch. *Phys. Rev.*, B52:10510, 1995.
106. C. R. A. Catlow and W. C. Mackrodt, editors. *Computer Simulation of Solids*, volume 166 of Lecture Notes in Physics. Springer, Berlin Heidelberg, New York, 1982.
107. C. R. A. Catlow, I. D. Faux and M. J. Norgett. *J. Phys. C: Solid State Phys.*, 5:415, 1976.
108. C. R. A. Catlow, I. D. Faux and M. J. Norgett. *J. Phys. C: Solid State Phys.*, 9:419, 1978.
109. M. Leslie. *Solid State Ionics*, 8:243, 1983.
110. J. D. Gale. GULP – General Utility Lattice Programme, Royal Institution, London, 1993.
111. W. C. Mackrodt. *J. Chem. Soc., Faraday Trans 2*, 85:541, 1989.
112. R. Pandey, X. Yang, J. M. Vail and J. Zuo. *Solid State Commun.*, 81:549, 1992.
113. J. H. Harding and A. H. Harker. *Phil. Mag.*, B51:119, 1985.
114. P. Saul, C. R. A. Catlow and J. Kendrick. *Phil. Mag.*, B51:107, 1985.
115. R. G. Gordon and Y. S. Kim. *J. Chem. Phys.*, 56:3122, 1972.
116. W. A. Harrison and R. Sokel. *J. Chem. Phys.*, 65:379, 1976.
117. W. C. Mackrodt and R. F. Stewart. *J. Phys. C: Solid State Phys.*, 12:431, 1979.
118. J. M. Recio, E. Francisco, M. Flórez and A. M. Pendás. *J. Phys.: Condens. Matter*, 5:4975, 1993.
119. R. Pandey, Jun Zuo and A. B. Kunz. *J. Mater. Res.*, 5:623, 1990.
120. R. W. Grimes. *Mol. Simulation*, 5:9, 1990.
121. A. Takada, C. R. A. Catlow and G. D. Price. *J. Phys.: Condens. Matter*, 7:8659, 1995.
122. G. V. Lewis and C. R. A. Catlow. *J. Phys. C: Solid State Phys.*, 18:1149, 1985.
123. M. Exner. private communication.
124. H. Donnerberg, S. M. Tomlinson, C. R. A. Catlow and O. F. Schirmer. *Phys. Rev.*, B40:11909, 1989.
125. S. Huzinaga, J. Andzelm, M. Klobukowski, E. Radzio-Andzelm, Y. Sakai and H. Tatewaki, editors. *Gaussian Basis Sets for Molecular Calculations.* Elsevier, Amsterdam, 1984.
126. T. H. Dunning and P. J. Hay. Gaussian basis sets for molecular calculations. In H. F. Schaefer (III), editor, *Methods of Electronic Structure Theory*, Chapter 1, page 1. Plenum Press, New York, 1977.
127. S. F. Boys and F. Bernardi. *Mol. Phys.*, 19:553, 1970.
128. R. W. Grimes, C. R. A. Catlow and A. M. Stoneham. *J. Phys.: Condens. Matter*, 1:7367, 1989.
129. C. R. A. Catlow and M. R. Hayns. *J. Phys. C: Solid State Phys.*, 5:L237, 1972.
130. M. J. Mehl, R. J. Hemley and L. L. Boyer. *Phys. Rev.*, B33:8685, 1986.
131. R. A. Jackson and C. R. A. Catlow. *Mol. Simulation*, 1:207, 1988.
132. K. Buse. *J. Opt. Soc. Am.*, B10:1266, 1993.
133. O. F. Schirmer, H.-J. Reyher and M. Wöhlecke. Characterization of point defects in photorefractive oxide crystals by paramagnetic resonance methods. In F. Agulló-López, editor, *Insulating Materials for Optoelectronics – New Developments.* World Scientific, Singapore, 1995.
134. M. S. Islam, D. J. Ilett and S. C. Parker. *J. Phys. Chem.*, 98:9637, 1994.
135. Z. Gong and R. E. Cohen. *Ferroelectrics*, 136:113, 1992.

136. R. Migoni, H. Bilz and D. Bäuerle. *Phys. Rev. Lett.*, 37:1155, 1976.
137. C. H. Perry, R. Currat, H. Buhay, R. M. Migoni, W. G. Stirling and J. D. Axe. *Phys. Rev.*, B39:8666, 1989.
138. H. Bilz, G. Benedek and A. Bussmann-Holder. *Phys. Rev.*, B35:4840, 1987.
139. A. Bussmann-Holder. *Ferroelectrics*, 136:27, 1992.
140. P. J. Edwardson. *Phys. Rev. Lett.*, 63:55, 1989.
141. R. E. Cohen and H. Krakauer. *Ferroelectrics*, 136:65, 1992.
142. R. E. Cohen. *Ferroelectrics*, 150:1, 1993.
143. H. Donnerberg and M. Exner. *Phys. Rev.*, B49:3746, 1994.
144. D. J. Singh and L. L. Boyer. *Ferroelectrics*, 136:95, 1992.
145. R. D. King-Smith and D. Vanderbilt. *Phys. Rev.*, B49:5828, 1994.
146. K. M. Rabe and U. V. Waghmare. *Ferroelectrics*, 136:147, 1992.
147. W. Zhong, D. Vanderbilt and K. M. Rabe. *Phys. Rev. Lett.*, 73:1861, 1994.
148. K. M. Rabe and J. D. Joannopoulos. *Phys. Rev. Lett.*, 59:570, 1987.
149. K. M. Rabe and J. D. Joannopoulos. *Phys. Rev.*, B36:6631, 1987.
150. D. J. Singh. *Phys. Rev.*, B52:12559, 1995.
151. A. V. Postnikov and G. Borstel. *Phys. Rev.*, B50:16403, 1994.
152. R. D. King-Smith and D. Vanderbilt. *Phys. Rev.*, B47:1651, 1993.
153. R. Resta. *Ferroelectrics*, 151:49, 1994.
154. M. V. Berry. *Proc. Roy. Soc. Lond.*, 392:45, 1984.
155. R. Resta, M. Posternak and A. Baldereschi. *Phys. Rev. Lett*, 70:1010, 1993.
156. W. Zhong, R. D. King-Smith and D. Vanderbilt. *Phys. Rev. Lett*, 72:3618, 1994.
157. F. Bernardini, V. Fiorentini and D. Vanderbilt. *cond-mat/9707252*, 1997.
158. G. V. Lewis and C. R. A. Catlow. *J. Phys. Chem. Solids*, 47:89, 1986.
159. F. M. Michel-Calendini and K. A. Müller. *Solid State Commun.*, 40:255, 1981.
160. F. M. Michel-Calendini and P. Moretti. *Phys. Rev.*, B27:763, 1983.
161. P. Moretti and F. M. Michel-Calendini. *Phys. Rev.*, B34:8538, 1986.
162. P. Moretti and F. M. Michel-Calendini. *J. Opt. Soc. Am.*, B5:1697, 1988.
163. M. Cherry, M. S. Islam, J. D. Gale and C. R. A. Catlow. *J. Phys. Chem.*, 99:14614, 1995.
164. J. Padilla and D. Vanderbilt. *Phys. Rev.*, B56:1625, 1997.
165. A. Mooradian, T. Jaegar and P. Stokseth. *Tunable Lasers and Applications.* Springer, Berlin Heidelberg, New York, 1976.
166. K. A. Müller, W. Berlinger and K. W. Blazey. *Solid State Commun.*, 61:21, 1987.
167. K. A. Müller. *Helv. Phys. Acta*, 59:874, 1986.
168. D. C. Rawlings and E. R. Davidson. *Chem. Phys. Lett.*, 98:424, 1983.
169. E. R. Davidson and D. W. Silver. *Chem. Phys. Lett.*, 52:403, 1977.
170. C. F. Bender and E. R. Davidson. *J. Phys. Chem.*, 70:2675, 1966.
171. E. R. Davidson. QCPE 23, Program No. 580, 1991.
172. M. Dupuis, J. D. Watts, H. O. Villar and G. J. B. Hurst. QCPE 23, Program No. 544, 1991.
173. R. D. Amos et al. Cambridge Analytic Derivatives Package (CADPAC) version 5.2, Cambridge, 1994.
174. R. S. Sinkovits and R. H. Bartram. Modification of HADES III, 1991.
175. H. Donnerberg. Modification of CASCADE, 1992.
176. R. D. Shannon and C. T. Prewitt. *Acta Cryst.*, B25:925, 1969.
177. M. J. L. Sangster. *J. Phys. C: Solid State Phys.*, 14:2889, 1981.
178. D. J. Groh, R. Pandey and J. M. Recio. *Phys. Rev.*, B50:14860, 1994.
179. G. Blasse, P. H. M. De Korte and A. Mackor. *J. Inorg. Nucl. Chem.*, 43:1499, 1981.
180. S. E. Stokowski and A. L. Schawlow. *Phys. Rev.*, 178:2, 1969.

181. S. E. Stokowski and A. L. Schawlow. *Phys. Rev.*, 178:457, 1969.
182. P. O. Löwdin. *Rev. Mod. Phys.*, 35:496, 1963.
183. P. S. Bagus and H. F. Schaefer III. *J. Chem. Phys.*, 56:224, 1972.
184. E. Possenriede, P. Jacobs, H. Kröse and O. F. Schirmer. *Appl. Phys.*, A55:73, 1992.
185. H. Donnerberg and A. Birkholz. *J. Phys.: Condens. Matter*, 7:327, 1995.
186. H. Donnerberg. *J. Phys.: Condens. Matter*, 7:L689, 1995.
187. H. Donnerberg, S. Többen and A. Birkholz. *J. Phys.: Condens. Matter*, 9:6359, 1997.
188. W. Känzig. *Phys. Rev.*, 99:1890, 1955.
189. P. W. Jacobs and E. A. Kotomin. *J. Phys.: Condens. Matter*, 4:7531, 1992.
190. G. Herzberg. *Molecular Spectra and Molecular Structure*, volume 1. Van Nostrand Reinhold Company, New York, 1950.
191. S. H. Vosko, L. Wilk and M. Nusair. *Can. J. Phys.*, 58:1200, 1980.
192. E. Possenriede, H. Kröse, T. Varnhorst, R. Scharfschwerdt and O. F. Schirmer. *Ferroelectrics*, 151:199, 1994.
193. A. M. Stoneham and M. J. L. Sangster. *Phil. Mag.*, B43:609, 1981.
194. E. Possenriede, P. Jacobs and O. F. Schirmer. *J. Phys.: Condens. Matter*, 4:4719, 1992.
195. J. Fink, N. Nücker, H. A. Romberg and J. C. Fuggle. *IBM J. Res. Develop.*, 33:372, 1989.
196. W. E. Pickett. *Rev. Mod. Phys.*, 61:433, 1989.
197. A. L. Shluger, E. A. Kotomin and L. N. Kantorovich. *J. Phys. C: Solid State Phys.*, 19:4183, 1986.
198. A. L. Shluger, E. N. Heifets, J. D. Gale and C. R. A. Catlow. *J. Phys.: Condens. Matter*, 4:5711, 1992.
199. J. Meng, P. Jena and J. M. Vail. *J. Phys.: Condens. Matter*, 2:10371, 1990.
200. L. Kantorovich, A. Stashans, E. Kotomin and P. W. M. Jacobs. *Int. J. Quant. Chem.*, 52:1177, 1994.
201. W. Kutzelnigg. *Einführung in die Theoretische Chemie*, volume 2. Verlag Chemie, Weinheim, New York, 1978.
202. H. Nakatsuji and H. Nakai. *Chem. Phys. Lett.*, 197:339, 1992.
203. R. Broer, A. B. Van Oosten and W. C. Nieuwpoort. *Reviews of Solid State Science*, 5:79, 1991.
204. B. Ya. Moǐzhes and S. G. Suprun. *Sov. Phys. Solid State*, 26:544, 1984.
205. F. A. Kröger and H. J. Vink. In F. Seitz and D. Turnbull, editors, *Solid State Physics*, volume 3. Academic Press, New York, 1956.
206. M. Exner, C. R. A. Catlow, H. Donnerberg and O. F. Schirmer. *J. Phys.: Condens. Matter*, 6:3379, 1994.
207. M. Exner. PhD thesis, University of Osnabrück, 1993.
208. U. van Stevendaal, K. Buse, S. Kämper, H. Hesse and E. Krätzig. *Appl. Phys. B*, 63:315, 1996.
209. S. A. Prosandeyev, A. V. Fisenko, A. I. Riabchinski and I. A. Osipenko. *J. Phys.: Condens. Matter*, 8:6705, 1996.
210. M. Tsukada, C. Satoko and H. Adachi. *J. Phys. Soc. Jap.*, 48:200, 1980.
211. R. Scharfschwerdt, A. Mazur, O. F. Schirmer, H. Hesse and S. Mendricks. *Phys. Rev.*, B54:15284, 1996.
212. A. V. Postnikov. private communication.
213. A. S. Nowick and D. S. Park. *Superior Conductors*. Plenum Press, New York, 1976.
214. M. E. Lines and A. M. Glass. *Principles and Applications of Ferroelectrics and Related Materials*. Clarendon Press, Oxford, 1977.

215. T. Neumann, G. Borstel, C. Scharfschwerdt and M. Neumann. *Phys. Rev.*, B46:10623, 1992.
216. Yong-Nian Xu, W. Y. Ching and R. H. French. *Ferroelectrics*, 111:23, 1990.
217. G. Godefroy and B. Jannot. *Key Engineering Materials*, 68:81, 1992.
218. C. Medrano and P. Günther. The photorefractive effect in crystals. In F. Agulló-López, editor, *Insulating Materials for Optoelectronics – New Developments*. World Scientific, Singapore, 1995.
219. B. E. Vugmeister and M. D. Glinchuk. *Rev. Mod. Phys.*, 62:993, 1990.
220. U. T. Höchli, K. Knorr and A. Loidl. *Adv. Phys.*, 39:405, 1990.
221. S. R. Andrews. *J. Phys. C: Solid State Phys.*, 18:1357, 1985.
222. H.-J. Reyher, B. Faust, M. Käding, H. Hesse, E. Ruža and M. Wöhlecke. *Phys. Rev.*, B51:6707, 1995.
223. C. Fischer, C. Auf der Horst, P. Voigt, S. Kapphan and J. Zhao. *Radiation Effects and Defects in Solids*, 136:85, 1995.
224. C. R. A. Catlow. *Proc. Roy. Soc. Lond.*, A353:533, 1977.
225. R. E. Cohen. *Nature*, 358:136, 1992.
226. R. E. Cohen. *Nature*, 362:213, 1993.
227. H. Thomann. *Ferroelectrics*, 73:183, 1987.
228. M. Exner, H. Donnerberg, C. R. A. Catlow and O. F. Schirmer. *Phys. Rev.*, B52:3930, 1995.
229. P. Pertosa and F. M. Michel-Calendini. *Phys. Rev.*, B17:2011, 1978.
230. G. O. Deputy and R. W. Vest. *J. Am. Ceram. Soc.*, 61:321, 1978.
231. W. K. Lee, A. S. Nowick and L. A. Boatner. Protonic conduction and nonstoichiometry in potassium tantalate crystals. In *Advances in Ceramics*, volume 23. The American Ceramic Society, Inc., Columbus (Ohio), 1987.
232. E. Wiesendanger. *Ferroelectrics*, 6:263, 1974.
233. O. Beck, D. Kollewe, A. Kling, W. Heiland and H. Hesse. *Nucl. Instr. Meth.*, B85:474, 1994.
234. D. M. Hannon. *Phys. Rev.*, 164:366, 1967.
235. B. E. Vugmeister and M. D. Glinchuk. *Sov. Phys. JETP*, 22:482, 1980.
236. S. G. Ingle and J. G. Dupare. *Phys. Rev.*, B45:2638, 1992.
237. D. J. Newman. *Adv. Phys.*, 20:197, 1971.
238. D. J. Newman and W. Urban. *Adv. Phys.*, 24:793, 1975.
239. E. Siegel and K. A. Müller. *Phys. Rev.*, B20:3587, 1979.
240. H. Arend. In V. Dvorak, A. Fouskova and P. Glogar, editors, *Proceedings of the International Meeting on Ferroelectricity*, page 231, 1966.
241. H. Ihrig. *J. Phys. C: Solid State Phys.*, 11:819, 1978.
242. P. Novak and I. Veltrusky. *phys. stat. sol. (b)*, 73:575, 1976.
243. K. A. Müller and W. Berlinger. *J. Phys. C: Solid State Phys.*, 16:6861, 1983.
244. M. Heming, S. Remme and G. Lehmann. *Ber. Bunsenges. Phys. Chem.*, 88:946, 1984.
245. E. Siegel and K. A. Müller. *Phys. Rev.*, B19:109, 1979.
246. H. Donnerberg, M. Exner and C. R. A. Catlow. *Phys. Rev.*, B47:14, 1993.
247. T. W. D. Farley, W. Hayes, S. Hull, M. T. Hutchings, R. Ward and M. Alba. *Solid State Ionics*, 28–30:189, 1988.
248. G. V. Lewis. PhD thesis, University College of London, 1984.
249. O. Hanske-Petitpierre, Y. Yacoby, J. Mustre de Leon, E. A. Stern and J. J. Rehr. *Phys. Rev.*, B44:6700, 1991.
250. M. Dubus, B. Daudin, B. Salce and L. A. Boatner. *Solid State Commun.*, 55:759, 1985.
251. G. E. Kugel, M. D. Fontana and W. Kress. *Phys. Rev.*, B35:813, 1987.
252. L. Foussadier, M. D. Fontana and W. Kress. *J. Phys.: Condens. Matter*, 8:1135, 1996.

253. M. G. Stachiotti and R. L. Migoni. *J. Phys.: Condens. Matter*, 2:4341, 1990.
254. A. V. Chadwick, K. W. Flack, J. H. Strange and J. Harding. *Solid State Ionics*, 28–30:185, 1988.
255. F. Borsa, U. T. Höchli, J. J. van der Klink and D. Rytz. *Phys. Rev. Lett.*, 45:1884, 1980.
256. J. J. van der Klink and S. N. Khanna. *Phys. Rev.*, B29:2415, 1984.
257. P. Voigt and S. Kapphan. *J. Phys. Chem. Solids*, 55:853, 1994.
258. V. Vikhnin, P. Voigt and S. Kapphan. *J. Phys.: Condens. Matter*, 7:7227, 1995.
259. R. I. Eglitis, A. V. Postnikov and G. Borstel. *Phys. Rev.*, B55:12976, 1997.
260. M. Stachiotti, R. Migoni, H.-M. Christen, J. Kohanoff and U. T. Höchli. *J. Phys.: Condens. Matter*, 6:4297, 1994.
261. R. I. Eglitis, E. A. Kotomin, A. V. Postnikov, N. E. Christensen and G. Borstel. *phys. stat. sol. (b) (to be published)*, 1998.
262. E. Possenriede, O. F. Schirmer, H. Donnerberg and B. Hellermann. *J. Phys.: Condens. Matter*, 1:7267, 1989.
263. H. Donnerberg. *Phys. Rev.*, B50:9053, 1994.
264. I. P. Bykov, M. D. Glinchuk, A. A. Karmazin and V. V. Laguta. *Sov. Phys. Solid State*, 25:2063, 1983.
265. Zhou Yi-Yang. *Phys. Rev.*, B42:917, 1990.
266. Zheng Wen-Chen. *Phys. Rev.*, B45:3156, 1992.
267. Gi-Guo Zhong, Jin Jiang and Zhong-Kang Wu. In *Proceedings of the 11th International Quantum Electronics Conference, IEEE Cat. No. CH1561-0*, page 631, 1980.
268. K. L. Sweeney, L. E. Halliburton, D. A. Bryan, R. R. Rice, R. Gerson and H. E. Tomaschke. *J. Appl. Phys.*, 57:1036, 1985.
269. F. Klose, M. Wöhlecke and S. Kapphan. *Ferroelectrics*, 92:181, 1989.
270. H. D. Megaw and C. N. W. Darlington. *Acta Cryst.*, A31:161, 1975.
271. I. B. Kituk. *Ukr. J. Phys.*, 35:1542, 1990.
272. L. Hafid and F. M. Michel-Calendini. *J. Phys. C: Solid State Phys.*, 19:2907, 1986.
273. F. M. Michel-Calendini et al. *Cryst. Latt. Def. and Amorph. Mat.*, 15, 1987.
274. Yuanwu Qiu. *J. Phys.: Condens. Matter*, 5:2041, 1993.
275. G. G. DeLeo, J. L. Dobson, M. F. Masters and L. H. Bonjack. *Phys. Rev.*, B37:8349, 1988.
276. F. M. Michel-Calendini, K. Bellafrouh and C. Daul. *Ferroelectrics*, 125:271, 1992.
277. F. M. Michel-Calendini, K. Bellafrouh and C. Daul. In *Proceedings of the XII International Conference on Defects in Insulating Materials*, volume 2, page 1184, 1992.
278. O. Gunnarsson and B. I. Lundqvist. *Phys. Rev.*, B13:4274, 1976.
279. H. Donnerberg, S. M. Tomlinson, C. R. A. Catlow and O. F. Schirmer. *Phys. Rev.*, B44:4877, 1991.
280. H. Donnerberg. *J. Solid State Chem.*, 123:208, 1996.
281. A. Kling, J. C. Soares and M. F. da Silva. Channeling Investigations of Oxide Materials for Optoelectronic Applications. In F. Agulló-López, editor, *Insulating Materials for Optoelectronics – New Developments*. World Scientific, Singapore, 1995.
282. S. C. Abrahams and P. Marsh. *Acta Cryst.*, B42:61, 1986.
283. G. E. Peterson and A. Carnevale. *J. Chem. Phys.*, 56:4848, 1972.
284. D. M. Smyth. In *Proceedings of the Sixth IEEE International Symposium on Applications of Ferroelectrics*, page 115, 1986.

285. A. Mehta, A. Navrotsky, N. Kumada and N. Kinomura. *J. Solid State Chem.*, 102:213, 1993.
286. N. Iyi, K. Kitamura, F. Izumi, J. K. Yamamoto, T. Hayashi, H. Asanoand and S. Kimura. *J. Solid State Chem.*, 101:340, 1992.
287. A. P. Wilkinson, A. K. Cheetham and R. H. Jarman. *J. Appl. Phys.*, 74:3080, 1993.
288. J. Blümel, E. Born and Th. Metzger. *J. Phys. Chem. Solids*, 55:589, 1994.
289. M. E. Twigg, D. N. Maher, S. Nakahara, T. T. Sheng and R. J. Holmes. *Appl. Phys. Lett.*, 50:501, 1987.
290. H. Müller and O. F. Schirmer. *Ferroelectrics*, 125:319, 1991.
291. K. Wulf, H. Müller, O. F. Schirmer and B. C. Grabmaier. *Radiation Effects and Defects in Solids*, 119–121:687, 1991.
292. G. I. Malovichko, V. G. Grachev, L. P. Yurchenko, V. Ya. Proshko, E. P. Kokanyan and V. T. Gabrielyan. *phys. stat. sol. (a)*, 133:K29, 1992.
293. A. Böker, H. Donnerberg, O. F. Schirmer and Feng Xiqi. *J. Phys.: Condens. Matter*, 2:6865, 1990.
294. G. Corradi, A. V. Chadwick, A. R. West, K. Cruickshank and M. Paul. *Radiation Effects and Defects in Solids*, 134:219, 1995.
295. G. I. Malovichko, V. G. Grachev, O. F. Schirmer and B. Faust. *J. Phys.: Condens. Matter*, 5:3971, 1993.
296. B. C. Grabmaier and F. Otto. *J. Crystal Growth*, 79:682, 1986.
297. B. C. Grabmaier and F. Otto. In *Proc. SPIE Int. Soc. Opt. Eng. (USA)*, volume 651, page 2, 1986.
298. N. Iyi, K. Kitamura, Y. Yajima and S. Kimura. *J. Solid State Chem.*, 118:148, 1995.
299. T. Volk, N. Rubinina and M. Wöhlecke. *J. Opt. Soc. Am. B*, 11:1681, 1994.
300. T. Volk, M. Wöhlecke, A. Reichert, F. Jermann and N. Rubinina. *Ferroelectrics Lett.*, 20:97, 1995.
301. U. Schlarb and K. Betzler. *Phys. Rev.*, B50:751, 1994.
302. T. Volk, M. Wöhlecke and H. Donnerberg. *Ferroelectrics Lett. (in press)*, 1996.
303. T. Volk, M. Wöhlecke, N. Rubinina, N. V. Razumovski, F. Jermann, C. Fischer and R. Böwer. *Appl. Phys.*, A60:217, 1995.
304. R. R. Neurgaonkar, W. K. Cory, J. R. Oliver, M. D. Ewbank and W. F. Hall. *Optical Engineering*, 26:392, 1987.
305. R. C. Baetzold. *Phys. Rev.*, B48:5789, 1993.
306. S. Haussühl, D. Mateika and W. Tolksdorf. *Z. Naturforschg.*, 31a:390, 1976.
307. J. W. Nielsen and E. F. Dearborn. *J. Phys. Chem. Solids*, 5:202, 1958.
308. K. Wagner, H. Lüttgemeier, W. Zinn, P. Novák, J. Englich, H. Dötsch and S. Sure. *Journal of Magnetism and Magnetic Materials*, 140–144:2107, 1995.
309. H. Donnerberg. unpublished.
310. R. Metselaar and M. A. H. Huyberts. *J. Phys. Chem. Solids*, 34:2257, 1973.
311. A. E. Paladino and E. A. Maguire. *J. Amer. Ceram. Soc.*, 53:98, 1970.
312. J. B. Mac Chesney and J. F. Potter. *J. Amer. Ceram. Soc.*, 48:534, 1965.
313. P. Hansen, K. Witter and W. Tolksdorf. *Phys. Rev.*, B27:6608, 1983.
314. P. K. Larsen and R. Metselaar. *J. Solid State Chem.*, 12:253, 1975.
315. R. C. Weast, editor. *Handbook of Chemistry and Physics*. CRC Press, Boca Raton (Florida), 63rd edition, 1982–83.
316. L. Schuh, R. Metselaar and C. R. A. Catlow. *J. Europ. Ceram Soc.*, 7:67, 1991.
317. N. Kimizuka and T. Katsura. *J. Solid State Chem.*, 13:176, 1975.
318. P. K. Larsen and R. Metselaar. *Phys. Rev.*, B14:2520, 1976.
319. H. Böttger and V. V. Bryksin. *Phys. Status Solidi*, 78:415, 1976.

320. P. Nagels. *The Hall Effect and its Applications.* Plenum Press, New York, 1980.
321. D. Emin. *Phys. Rev.*, B3:1321, 1971.
322. M. J. Norgett and A. M. Stoneham. *J. Phys. C: Solid State Phys.*, 6:229, 1973.
323. T. Holstein. *Ann. Phys. (New York)*, 8:325, 1959.
324. S. H. Wemple, S. L. Blank, J. A. Seman and W. A. Biolsi. *Phys. Rev.*, B9:2134, 1974.
325. S. P. Kowalczyk, F. R. McFeely, L. Ley, V. T. Gritsyna and D. A. Shirley. *Solid State Commun.*, 23:161, 1977.
326. P. M. Grant and W. Ruppel. *Solid State Commun.*, 5:543, 1967.
327. R. A. Jackson. unpublished.
328. H. Donnerberg. unpublished.
329. V. Lupei, L. Lou, G. Boulon and A. Lupei. *J. Phys.: Condens. Matter*, 5:L35, 1993.
330. L. Suchow, M. Kokta and V. J. Flynn. *J. Solid State Chem.*, 2:137, 1970.
331. I. Nowik. *Phys. Rev.*, 171:550, 1968.
332. A. Rosencwaig. *Canad. J. Phys.*, 48:2868, 1970.
333. R. Gonano, E. Hunt and H. Meyer. *Phys. Rev.*, 156:521, 1967.
334. J. M. Robertson, J. C. Verplanke, S. Wittekoek, P. F. Bongers, M. Jansen and A. Op den Buis. *Appl. Phys.*, 6:353, 1975.
335. J. E. Davies, E. A. Giess, J. D. Kuptsis and W. Reuter. *J. Cryst. Growth*, 36:191, 1976.
336. P. Paroli and S. Geller. *J. Appl. Phys.*, 48:1364, 1977.
337. Landolt-Börnstein. *Zahlenwerte und Funktionen aus Physik, Chemie, Astronomie, Geophysik, Technik*, volume I/1. Springer, Berlin, Göttingen, Heidelberg, sixth edition, 1951.

Index

Springer Tracts in Modern Physics

133 **Matter at High Densities in Astrophysics**
Compact Stars and the Equation of State
In Honor of Friedrich Hund's 100th Birthday
By H. Riffert, H. Müther, H. Herold, and H. Ruder 1996. 86 figs. XIV, 278 pages

134 **Fermi Surfaces of Low-Dimensional Organic Metals and Superconductors**
By J. Wosnitza 1996. 88 figs. VIII, 172 pages

135 **From Coherent Tunneling to Relaxation**
Dissipative Quantum Dynamics of Interacting Defects
By A. Würger 1996. 51 figs. VIII, 216 pages

136 **Optical Properties of Semiconductor Quantum Dots**
By U. Woggon 1997. 126 figs. VIII, 252 pages

137 **The Mott Metal-Insulator Transition**
Models and Methods
By F. Gebhard 1997. 38 figs. XVI, 322 pages

138 **The Partonic Structure of the Photon**
Photoproduction at the Lepton-Proton Collider HERA
By M. Erdmann 1997. 54 figs. X, 118 pages

139 **Aharonov–Bohm and Other Cyclic Phenomena**
By J. Hamilton 1997. 34 figs. X, 186 pages

140 **Exclusive Production of Neutral Vector Mesons at the Electron-Proton Collider HERA**
By J. A. Crittenden 1997. 34 figs. VIII, 108 pages

141 **Disordered Alloys**
Diffusive Scattering and Monte Carlo Simulations
By W. Schweika 1998. 48 figs. X, 126 pages

142 **Phonon Raman Scattering in Semiconductors, Quantum Wells and Superlattices**
Basic Results and Applications
By T. Ruf 1998. 143 figs. VIII, 252 pages

143 **Femtosecond Real-Time Spectroscopy of Small Molecules and Clusters**
By E. Schreiber 1998. 131 figs. XII, 212 pages

144 **New Aspects of Electromagnetic and Acoustic Wave Diffusion**
By POAN Research Group 1998. 31 figs. IX, 117 pages

145 **Handbook of Feynman Path Integrals**
By C. Grosche and F. Steiner 1998. X, 449 pages

146 **Low-Energy Ion Irradiation of Solid Surfaces**
By H. Gnaser 1999. 93 figs. VIII, 293 pages

147 **Dispersion, Complex Analysis and Optical Spectroscopy**
By K.-E. Peiponen, E.M. Vartiainen, and T. Asakura 1999. 46 figs. VIII, 130 pages

148 **X-Ray Scattering from Soft-Matter Thin Films**
Materials Science and Basic Research
By M. Tolan 1999. 98 figs. IX, 197 pages

149 **High-Resolution X-Ray Scattering from Thin Films and Multilayers**
By V. Holý, U. Pietsch, and T. Baumbach 1999. 159 figs. XI, 255 pages

150 **QCD at HERA**
The Hadronic Final State in Deep Inelastic Scattering
By M. Kuhlen 1999. 99 figs. X, 172 pages

151 **Atomic Simulation of Electrooptic and Magnetooptic Oxide Materials**
By H. Donnerberg 1999. 45 figs. VIII, 205 pages

152 **Thermocapillary Convection in Models of Crystal Growth**
By H. Kuhlmann 1999. 101 figs. XVIII, 224 pages

Springer and the environment

Printing: Mercedesdruck, Berlin
Binding: Buchbinderei Lüderitz & Bauer, Berlin